THE WOMAN WHO DECIDED TO DIE

*Challenges and Choices at the
Edges of Medicine*

Ronald Munson

OXFORD
UNIVERSITY PRESS

2009

OXFORD
UNIVERSITY PRESS

Oxford University Press, Inc., publishes works that further
Oxford University's objective of excellence
in research, scholarship, and education.

Oxford New York
Auckland Cape Town Dar es Salaam Hong Kong Karachi
Kuala Lumpur Madrid Melbourne Mexico City Nairobi
New Delhi Shanghai Taipei Toronto

With offices in
Argentina Austria Brazil Chile Czech Republic France Greece
Guatemala Hungary Italy Japan Poland Portugal Singapore
South Korea Switzerland Thailand Turkey Ukraine Vietnam

Published by Oxford University Press, Inc.
198 Madison Avenue, New York, New York 10016

www.oup.com

Oxford is a registered trademark of Oxford University Press

Library of Congress Cataloging-in-Publication Data
Munson, Ronald, [date]
The woman who decided to die : challenges and choices
at the edges of medicine / Ronald Munson.
p. ; cm.
Includes bibliographical references.
ISBN 978-0-19-533101-1
1. Medical ethics—Case studies. I. Title.
[DNLM: 1. Bioethical Issues. 2. Ethics, Medical. WB 60 M969w 2009]
R724.M874 2009
174.2—dc22   2008026796

1 3 5 7 9 8 6 4 2

Printed in the United States of America
on acid-free paper

TO MIRIAM

Donec gratus eram tibi.
—Horace, Odes III (ix)

Contents

Prologue

I think the case can be made that the world of American medicine changed decisively at two in the morning on April 14, 1975, when Julie Quinlan of Landing, New Jersey, was awakened by a telephone call. She was crying when she hung up the phone. "Karen is very sick," she said to her husband, Joseph. "She's unconscious, and we have to go to Newton Hospital right away."

The Quinlans thought their twenty-one-year-old adopted daughter must have been in an automobile accident, but at the hospital an ICU doctor informed them otherwise. Karen was in a comatose state for unknown reasons and was being given oxygen through a mask taped over her nose and mouth. It wasn't clear that she would live.

Karen had been brought to the hospital by two friends who had been with her at a birthday party. After a few drinks, she had started to pass out, and her friends decided she must be drunk and put her to bed. When they checked on her fifteen or twenty minutes later, they discovered that she wasn't breathing. They gave her mouth-to-mouth resuscitation, then rushed her to the nearest hospital.

Blood and urine tests indicated that Karen hadn't consumed what would usually be considered a dangerous amount of alcohol. But she had been on a very strict diet and had eaten nothing that day before having two gin-and-tonics at the party. The tests also detected the presence in her blood of .6 milligrams of aspirin combined with the tranquilizer Valium. Two milligrams would have been toxic, five lethal. So why she had stopped breathing was a mystery, yet it was during the brief period she was alone in the bedroom that part of her brain died, apparently from oxygen depletion.

About a week after Karen failed to regain consciousness, her parents moved her to St. Clare's Hospital, which was larger and had more advanced testing and life-support facilities. Dr. Robert J. Morse, a neurologist, and Dr. Arshad Javed, a pulmonary internist, became her physicians. They conducted additional studies to rule out several possible causes of her coma, including trauma, infection, and a brain tumor. The neurological tests and X-ray studies confirmed, however, that Karen's brain had undergone extensive damage.

NO LONGER THE SAME

The Quinlans were hopeful during Karen's early days in the hospital. Her eyes opened and closed, and her mother and her nineteen-year-old sister, Mary Ellen, thought they detected signs that Karen recognized them. Her father and her eighteen-year-old brother, John, also thought they noticed signs that Karen was aware of her surroundings.

Then Karen's condition began to deteriorate. Her weight dropped from 120 pounds to 70. Her body gradually began to draw in upon itself, contracting into a rigid fetal position. Her five-foot-two-inch frame curved into a shape hardly longer than three feet. To make sure Karen got enough oxygen, her doctors hooked her up to an MA-1 ventilator, a machine that pumped air into her lungs through a tube in her throat. She was fed liquids through a nasogastric tube, and an IV line dripped in fluids to keep her hydrated.

By early July, Karen's physicians and her mother, sister, and brother had come to believe that Karen would never regain consciousness. Only her father continued to think it might be possible. But when Joseph Quinlan mentioned to Dr. Morse some encouraging sign he thought he'd noticed, Dr. Morse told him, "Even if God did perform a miracle so that Karen would live, her damage is so extensive she would spend the rest of her life in an institution."

Her father realized then that Karen would never again be the way he remembered her. His little girl was gone. He came to accept Mary Ellen's view that "Karen would never want to be kept alive on machines like this. She would hate this."

THE PRIEST AND THE LAWYERS

The Quinlans' parish priest, Father Thomas Trapasso, assured the Quinlans that the ethical principles of the Roman Catholic Church didn't require the continuation of extraordinary measures to support a life after any realistic hope of recovery was gone.

"Am I playing God?" Mr. Quinlan asked the priest. He was still debating in his own mind the implications of asking the doctors to discontinue Karen's treatment.

"God has made the decision that Karen is going to die," Father Thomas told him. "You're just agreeing with God's decision, that's all."

On July 31, 1975, after Karen had been in a coma for three and a half months, Julie and Joseph Quinlan gave Drs. Morse and Jared their permission to take their daughter off the ventilator. They signed a letter authorizing the discontinuance of all extraordinary procedures and absolving the hospital from all legal liability that might result from Karen's death.

"I think you have come to the right decision," Dr. Morse told them.

But the next morning Dr. Morse called Joseph Quinlan and said, "I have a moral problem about what we agreed on last night. I feel I have to consult somebody else and see how he feels about it." The next day, Dr. Morse called again. "I find I will not do it," he said. "And I've informed the administrator at the hospital that I will not do it."

The Quinlans were upset and bewildered by the change in Dr. Morse. They talked with the hospital's attorney, and he told them that, because Karen was over twenty-one, they were no longer her legal guardians. They would have to go to court and be appointed to guardianship. After that, they would have the standing to ask the hospital to remove Karen from the respirator, and the hospital might or might not comply.

Joseph Quinlan consulted attorney Paul Armstrong. Because Karen was an adult without income, Mr. Quinlan explained, Medicare was paying the $450 a day it cost to keep her alive. The Quinlans thus had no financial motive in asking that she be removed from the ventilator. Joseph said his belief that Karen should be allowed to die rested on his conviction that it was God's will. His family agreed with him, and to carry out what they all wanted to happen, he sought to be appointed as Karen's guardian.

THE COURTS

Paul Armstrong filed a plea with Judge Robert Muir of the New Jersey Superior Court on September 12, 1975. He requested that Joseph Quinlan be appointed Karen's guardian so that he would have "the express power of authorizing the discontinuance of all extraordinary means of sustaining her life."

Armstrong argued the case on three constitutional grounds. First, he claimed (following the lead of the 1973 U.S. Supreme Court decision in *Roe v. Wade*) that there is an implicit right to privacy guaranteed by the Constitution and that this right permits individuals or others acting for them to terminate the use of extraordinary medical measures, even when death may result. This right holds, Armstrong said, unless there are compelling state interests that set it aside.

Second, Armstrong argued that the First Amendment guarantee of religious freedom extended to the Quinlan case. If the court didn't allow them to act in accordance with the doctrines of their church, their religious liberty would be infringed.

Finally, Armstrong appealed to the "cruel and unusual punishment" clause of the Eighth Amendment. He claimed that "for the state to require that Karen Quinlan be kept alive, against her will and the will of her family, after the dignity, beauty, promise and meaning of earthly life have vanished, is cruel and unusual punishment."

Julie and Mary Ellen Quinlan and one of Karen's friends testified that Karen had often talked about not wanting to be kept alive by machines. A neurologist called as an expert witness testified that Karen was in a "chronic vegetative state" and wasn't likely ever to regain consciousness.

Doctors testifying for St. Clare's Hospital, as well as Karen's own physicians, said they agreed with this conclusion. Even so, the hospital's attorney argued, Karen's brain showed patterns of electrical activity, and she still had a discernible pulse. Thus, because she couldn't be considered dead by legal or medical criteria, taking her off the ventilator would kill her.

On November 10, Judge Muir ruled against Joseph Quinlan. He praised Mr. Quinlan's character and concern, but he decided that Mr. Quinlan's anguish over his daughter might cloud his judgment about her welfare, so he shouldn't be made her guardian. Also, Judge Muir said, because Karen was still medically and legally alive, "the Court should not authorize termination of the respirator. To do so would be homicide and an act of euthanasia."

APPEAL

Paul Armstrong immediately filed an appeal with the New Jersey Supreme Court. On January 26, 1976, the court convened to hear arguments, and Armstrong made substantially the same case as before. This time the court's ruling was favorable. The court agreed that Joseph Quinlan could assert a right of privacy on Karen's behalf and that whatever he decided was in her best interest should be accepted by society. The court also set aside any criminal liability for removing the ventilator, stating that if this act resulted in death, it wouldn't be homicide, and even if it were homicide, it would not be an unlawful action.

Finally, the court stated, if Karen's doctors believed she would never emerge from her coma, they should consult an ethics committee that should be established by St. Clare's Hospital. If the committee accepted their prognosis, then the ventilator could be removed. If Karen's present physicians were then unwilling to take her off the respirator, Mr. Quinlan was free to find a physician who would.

NOT IN THIS HOSPITAL

Six weeks after the court decision, Karen's ventilator still hadn't been removed. Indeed, a second machine, one for controlling body temperature, had been added.

Joseph Quinlan met with Drs. Morse and Jared and demanded that they take Karen off the ventilator, and they agreed they would "wean" her off the machine. The Quinlans expected Karen to die once she was off the machine, but she was soon breathing on her own without mechanical assistance.

Dr. Morse and representatives of St. Clare's Hospital then made it clear to Mr. Quinlan that they were determined not to let Karen die while she was under their care. They moved her out of the ICU and into a private room, but the room was next door to the ICU. They announced that they intended to put her back on the ventilator at the first sign of any breathing difficulty.

Because Karen was still alive and breathing without assistance, the Quinlans were forced to began searching for a chronic-care hospital that would accept her as a patient. More than twenty institutions turned them away, and the physicians they talked to expressed great reluctance to become involved in such a controversial case. Finally, one physician, Dr. Joseph Fennelly, volunteered to take care of Karen, and on June 9 she was moved from St. Clare's Hospital to the Morris View Nursing Home.

TEN YEARS

Karen Quinlan continued to breathe on her own. She was given high-nutrient feedings through a surgically implanted gastric tube, and she received regular doses of antibiotics to ward off bacterial infections. Although still comatose, she was more active during some periods than others, grimacing and grunting and making reflexive responses to touch and sound. Her parents continued to visit her regularly.

On June 11, 1985, at 7:01 in the evening, ten years after she had lapsed into a coma at her friend's birthday party, Karen Quinlan finally died. She was thirty-one years old.

To most people in 1975, the case of Karen Quinlan seemed both astonishing and disturbing, and the whole country followed reports of the medical, moral, and legal issues as months passed and Karen neither died nor recovered consciousness. Most of us had never heard the phrase *persistent vegetative state*, which the neurologists used to describe her condition, and we were puzzled about its implications.

If Karen was trapped in some indefinite, ambiguous territory between life and death, how should we treat her? Were we duty-bound to keep her body functioning, even when she had no hope of recovery? The Quinlan's priest didn't think so, but what were his reasons? Maybe we should stop giving her medical care and let her die? Or maybe, shocking even to consider, we should end her life painlessly by administering a lethal drug?

In hearing about the dilemma facing Karen Quinlan's family, we all—because we and our loved ones were all potential patients—began to realize something that doctors had discovered in medical school: contemporary medicine isn't only about wonder drugs, surgical miracles, and fantastic new machines. Rather, medicine now had acquired a darker side. Not only do people die, but before that occurs, troubling questions can pop up unexpectedly like the horror-house skeletons at a carnival side-show.

People began to realize, as they followed the Quinlan story, that the new technologies for keeping patients alive—the IV lines, feeding-tubes, and mechanical ventilators—could also trap them in the shadowy realm between life and death.

We sensed at the time that anyone, given medicine's new and considerable powers, could end up like Karen, and many people were frantic to protect themselves from becoming prisoners in their own inert and permanently unconscious bodies. Brian Clark's 1978 play *Whose*

Life Is It Anyway? captured the anxiety triggered by the Quinlan case. Ken Harrison, the main character, is a sculptor rendered paralyzed from the neck down by a car accident. Although he can think and speak, he can't continue his work, and he wants the hospital staff to help him die. When they refuse, he goes to court to plead his case. That the play was a much-discussed Broadway hit suggests it expressed a worry shared by many.

Additional court decisions and criticisms by consumer advocates soon led to changes in the policies of hospitals. Many people insisted that patients be allowed to make their own end-of-life decisions and that doctors not be permitted to turn them into high-tech zombies, yet others argued that patients ought to be kept on life-support indefinitely, because it was always possible that they might recover consciousness.

(The furious battle over the fate of Terry Schiavo in 2005, another woman in a persistent vegetative state kept alive by a ventilator, erupted exactly thirty years after the Quinlan case became the focus of public debate. This suggests that the wheel of the controversy is now turning through another cycle.)

GESTALT SWITCH

I point to the Quinlan case as a pivotal moment of change in the way medicine is practiced because it was substantially responsible for two consequences. First, the case forced our society to recognize that medicine had acquired new and effective ways of controlling life and death that we knew little about. Most us had no idea that someone's life could be extended indefinitely by the use of drugs, IV hydration, diets fed through implanted gastric tubes, and other modes of life-support technology. (That Karen Quinlan went on living for ten years served as a continuing reminder of the power of that technology.)

The recognition of this possibility prompted us to start paying attention to some of the changes that had taken place in medicine. Soon we were hearing more about such comparatively new treatments as kidney dialysis, organ transplantation, open-heart surgery, and chemotherapy, and the more we learned about them, the more we began to realize that they too raised troubling questions: Should everybody needing dialysis qualify for it? Should we allow kidneys to be sold? Should we permit surgeons to perform novel and risky operations on children? Can a wife refuse treatment on behalf of her unconscious husband?

Second, when we began thinking about Karen Quinlan and euthanasia, we all began to ask, "Whose life is it anyway?" We realized that we no longer wanted to rely solely on doctors to make important medical decisions on our behalf. We began to insist that when we become patients, we don't surrender our autonomy. We want and need our physicians to provide us with information about our medical problems and tell us about the advantages and disadvantages of the treatment options. Ultimately, though, we want to make the final decision about what is done to us.

These two changes meant that our society was redefining the role of the patient. A patient was no longer merely the recipient of actions performed by the doctor, but someone who is given information by the doctor and then participates in the decision-making process. This amounted to a gestalt switch in the way our society viewed clinical medicine. The paternalistic idea behind "doctor's orders" and "for your own good" was replaced by the notion of informed consent, and when this happened, collaborative decision-making became the model for the doctor-patient relationship.

PUBLIC DISCUSSIONS, PERSONAL CHOICES

Viewing the doctor-patient relationship in this new way helped bring into the open ethical issues that had been discussed, if at all, only in the backrooms of medicine. Doctors treating newborns with severe impairments, for example, were accustomed to taking sole responsibility for deciding whether to end treatment and let the child die. Now parents were expected to help make that decision, and the conditions that would justify withholding treatment became a matter of often-fractious dispute. Abortion had been a volatile and divisive issue even before the Quinlan case, and it, along with euthanasia, continued to be argued about in terms of personal freedom.

The most significant change in medicine as a result of the Karen Quinlan case was the increase in the number of medical topics publicly debated, as research advanced and medicine developed new therapies and faced new disorders. The hottest topic of the 1970s was whether to risk research into recombinant-DNA technology, but in the early 1980s that issue was eclipsed by the more pressing problem of AIDS. Given the explosive spread of HIV infection, and with tens of thousands of people dying of AIDS each year, we faced a crisis that involved a volatile mixture of sex, infectious disease, threats to public health, lifestyle choices, privacy, personal responsibility, insurance, religion, drug testing, and politics.

The public—all of us as individuals—grew more and more accustomed to debating the ethical and social aspects of medical matters. Even a small sample of topics hints at the increasing sophistication of the associated issues: in vitro fertilization, egg donation, and frozen embryos; surrogate motherhood; heart and liver transplants; genetic testing and screening for disease; involuntary sterilization; psychoactive drugs; bone-marrow transplants; gene therapy; human-subjects research; cloning; and human embryonic stem cells.

Although little noticed, behind the public debates about policies, at the edges of medicine, were an increasing number of cases in which people were forced to make difficult, often agonizing, personal decisions. We had insisted that we no longer wanted our doctors to treat us as children, withhold the truth from us, and do whatever they decided was in our best interest. We wanted to take control of our own lives. Thus, the need to make a life-altering decision could suddenly erupt within the context of someone's particular medical circumstances and problems.

The need to decide was the skeleton that popped up to startle and frighten individuals: Should I agree to have a risky but potentially life-saving treatment? Should I ask my mother's doctor to remove her life support? Do I want to know the results of a blood test that might show I'm likely to develop Alzheimer's? Do I want to know if I'm infected with HIV? Should I agree to donate a liver segment to help my stepson? Should I consent to take part in a gene-therapy experiment? Can I ask my mentally retarded sister to donate a kidney to me?

And on and on.

MEDICAL ETHICISTS: PROFESSIONAL WORRIERS

As questions like these emerged within medicine, physicians realized they were facing issues they hadn't been trained to deal with in medical school. Many began to publish journal articles discussing their dilemmas, and soon philosophers trained in ethical theory were exchanging ideas with them. The doctors had to learn about ethical principles and arguments to defend their views, and the ethicists had to learn about the realities of medical diagnosis and treatment.

Medical ethics soon emerged as a subspecialty of philosophy. Ethicists could be useful people for physicians to consult about morally complex situations, and the role of medical ethicist became a recognized one at hospitals. Discussions were at first informal, but eventually getting an "ethics consult" became a standard practice. The role of a medical ethicist even now is only to clarify issues and offer opinions, rather than to make decisions rightfully made only by patients or their doctors. Some physicians became full-time medical ethicists, and several medical schools established programs to train MDs who are also medical ethicists. Most MD medical ethicists specialize in *clinical ethics*, the part of medical ethics that deals with problems associated with treating patients in the hospital.

Philosophers still write about the issues of medical ethics, even though most aren't on the clinical firing line. They have become *bioethicists*, expanding their concerns from the problems of patients to the broad range of value issues about medicine, biology, and society. Often they identify and try to resolve questions before they become practical problems. Thus, bioethicists were discussing cloning humans a decade before Dolly the sheep was born in 1996. They serve as what the writer Gina Kolata once called "professional worriers."

CASES IN THIS BOOK

My education and experiences as a medical ethicist (or now bioethicist) paralleled the dawning recognition of the new ethical and social problems of medicine. Thus, I was often present at the back of the hospital room or at the foot the elevated bed after a patient or family member had talked over matters with a doctor, then was forced to make a decision of great personal and moral importance. I was never in a position in which I was expected to make a decision, but I was sometimes asked for advice by those who had to make one. I was also more often asked for my views informally than formally.

The first part of each chapter of the ten that form the core of this book is devoted to telling the stories of patients, physicians, and families caught up in private, deadly serious medical dramas. One of my aims in presenting these cases is to bring back personal reports about the circumstances and challenges faced by the inhabitants of a country that even now is not especially well known—the country populated by the sick and injured, those who love them, and those who try to help them.

Too often, I think, we (ethicists, physicians, and society at large) discuss issues in medical ethics in ways that are so theoretical and impoverished that the human elements that give them urgency and significance are drained away. The result is that the emotional dynamics and complexities of situations are also eliminated. Often the ethical question is presented for analysis in the same terse formula used to state a diagnostic problem in medicine. Thus, the medical formula "A twenty-year-old male presented with a complaint of severe fatigue, easy bruising, and unhealing wounds" serves as the model for stating an ethical question: "A young man with metastatic cancer who has failed the standard therapies is faced with the choice of enrolling in a Phase I trial of an unproven chemotherapeutic drug."

Nothing is wrong about stating a moral problem abstractly, but for those who aren't familiar with clinical (bedside) medicine and the circumstances under which patients, doctors, and families must make choices, the abstract form fails to capture how very hard making decisions can be. It's one thing to talk about a twenty-year-old male and quite another to talk about Tom Hopkins's only child. Even though the problem is identical, the way those involved think about it and arrive at a decision differs from case to case.

Because I want to evoke such circumstances and complexities, I decided to present the cases as narrative accounts. I present them as possessing their own integrity and having an inherent interest and value, as found fragments of human experience. Thus, the cases are intended to evoke an emotional response, raise questions about human relationships, and offer a certain aesthetic satisfaction. My aim is to make the cases illuminate the people involved and the hard decisions they must make when confronted by trauma, disease, and the threat of imminent death. To enhance my chances of capturing the responses of people in critical medical circumstances, I use the techniques of personal narration to offer a specific point of view on events and people, employ precise descriptions to characterize actions and set scenes, and rely on dialogue to reveal character and attitude and move the case along to its conclusion.

Although I use narrative techniques to make the cases "read like fiction," the cases are based on ones I recall, but not ones I recorded. Thus, they have the character and qualities of remembered experiences. We recount to our friends what people said and did at the time of the event we're recalling, but no one believes we're quoting dialogue from a transcript or consulting contemporaneous notes. (In this respect, perhaps Wright Morris was right in saying that "Anything that passes through memory is fiction.")

The events I recount took place over many years and in many locations. Roughly, the time is from 1975 (the date of the Quinlan case) to the present, and the events occurred at several academic medical centers (Harvard, Johns Hopkins, University of California-San Diego, and Washington University), as well as at meetings of various divisions of the National Institutes of Health. I've taken care, however, to erase features that might identify the people or the places. I want to avoid violating confidentiality or even causing embarrassment. The truths in these accounts are thus (as philosophers say) always the essential ones, but not necessarily the accidental ones.

The cases all present problems that lie at the edges of medicine. These are the places where the magic of medicine often fails, treatments are unreliable, hope is uncertain, and death is a threatening outcome. The edges of medicine are where we get the sharpest pictures of people coping with helplessness, mortality, and doubts about how to act. I intend, as I said, for the cases to possess their own integrity and worth, but I also intend for them to raise questions about human relationships, reveal values in conflict, and evoke intellectual, as well as emotional, responses. I hope readers will care about these people in trouble, because they are us, and I hope readers will care about their problems, because they are ours.

ISSUES IN CONTEXT

My second aim in opening each chapter with a case is to present ethical issues in the contexts in which they arise. Thus, after each case presentation, I make my own move toward abstraction. I go beyond or behind the concrete features of the case to focus on the ethical quandary the central actors in the personal medical drama find themselves grappling with.

I don't, however, try to outline the broad issues that the case may also represent. Thus, for example, I don't use a particular case to discuss euthanasia in general: Definitions of, Types of, Distinct from homicide?, Kind of suicide? Ever justified?, and so on. Rather, I isolate, state, and explain ways to resolve in a justifiable fashion the specific ethical issue faced by those who must make the central moral decision.

If someone writes to "The Ethicist" columnist in the *New York Times* and asks, "Is it okay for me to take a plasma TV out of my employer's warehouse without paying for it, because he lied to me about giving me a raise?" that person wants an answer to the question, not a lecture on ethical theory or a comprehensive review of the circumstances in which stealing might be morally legitimate.

Similarly, I take it for granted that when someone asks, "Was it right for me to tell the doctor to let my husband die?" that's the question she wants answered. This is the kind of question I try to answer in the chapters, and I try to make the answers relevant, direct, and justified.

Yet I also try not to give such a terse answer that a question can't be seen as arising in the broader context within which patients and doctors make decisions. I believe, for example, that a discussion of the duty of a psychiatrist to respect a patient's confidentiality must address the question of what the psychiatrist should do when his patient threatens to harm someone. The best illustration of this situation is in the *Tarasoff* case, and I describe the incident, even though it's not directly relevant to the case I present. I try to walk the line between saying too little and saying too much in my comments on all the cases, and I hope I haven't strayed over the border in either direction too often.

The focus of my discussion in each case, I said earlier, is on the central ethical issue the doctors, patients, or family struggle to resolve. I don't want to create the impression, however, that every case raises only a single issue. Issues in medical ethics, like troubles in life, come not in single files but in battalions. If we stand back from almost any case, we can ask a large number of ethical questions, but those involved in the case usually grapple with only the issue that is most important to them. This doesn't mean they may not be aware of other issues, but they can't pay attention to everything at once and so set priorities. (People feverishly trying to shingle the roof before the rain comes also know their house needs painting.)

Suppose, to take an abstract case, the parents of a gravely injured grown son have been told by his doctors that he will never recover and that intensive care is necessary to keep his body functioning. In the abstract, we see a swarm of ethical and social issues hovering around this situation. Here, in no particular order, are a few categories of specific questions:

Who should pay? If the patient is uninsured, should the parents be asked to pay for his care? If he has no insurance and no assets, would the hospital be justified in ending his care if his parents aren't willing to pay the bills?

Is the patient dead? What if he has permanently lost all higher brain functions? What if he has lost all brain functions, but machines and drugs keep his body ticking away? Is it all right to remove his life support if either condition is satisfied?

Who should decide about terminating treatment? Is it legitimate for the parents to substitute their judgment for their son's and take him off life support? What if the parents know what he would want—and it's not what they want to do? Whose opinion should rule?

What about transplant organs? Can the patient's organs be removed once he has been declared dead by neurological criteria, even though his heart is still beating? Should the parents be able be able to sell their son's organs? What rules should govern who gets the organs?

Real people struggle with making decisions within the constraints imposed by medical circumstances, hospital policies, their own preferences, and the law. Thus, actual parents, in a situation like our abstract one, are unlikely to deliberate about (or even think of) most of the issues that prompt ethicists to worry. Real parents may wrestle primarily with the question of whether taking their son off life support is the best thing they can do for him. They may not be concerned in the least with whether a diagnosis of brain death means he's "really" dead. Nor may they think of their son as a potential organ donor, much less face the question of whether it's all right to sell his kidneys.

Had I taken an abstract approach in this book, I could have used only one case to illustrate any number of ethical issues. We can't quite see all the medical-ethics universe in the grain of sand that is a single case, but we can come close. What is left out, as I suggested earlier, is any appreciation of the fears, confusions, and uncertainties of people forced to make a decision when what is at stake is nothing less than their own life or the life of someone they love. These are elements I hope to capture, even if this means focusing on one issue in each case, even when many other issues could be raised.

PRINCIPLES

We live in a much more medically sophisticated society than we did at the time of the Quinlan case, so most people already have ideas about how questions in medical ethics should be answered. Even so, they rarely are able to articulate reasons for their views. In suggesting answers to the questions the cases raise, I make a point of offering reasons in support of them. They are reasons in terms of moral principles accepted by almost everyone in our society. Even though some may think the principles follow from an ethical theory and others believe they are justified by religion, the principles themselves aren't at issue. Who doesn't believe it's prima facie wrong to kill innocent people? Who denies that we should recognize the autonomy of people who are of sound mind?

The answers to ethical questions rarely (if ever) possess the sort of demonstrative certainty that we associate with mathematics and science. Even though I give reasoned answers, I don't expect everyone to agree with my conclusions. The cases raise significant and timely problems, so readers don't have to accept my solutions in order to appreciate the debate.

CODA

The private dramas played out at the edges of medicine have always fascinated me. The action is usually driven by the need to make crucial decisions in the face of inexorable forces. These dramas are often personal tragedies, filled with anguish and misery, yet some are thrilling tales of death averted by some medical *deus ex machina*. Humor is rare, yet irony abounds.

People like me, who think and write about questions of moral legitimacy and responsibility in medicine, tend to face these questions head-on and deal with them in an impersonal, no-nonsense, analytic way. This is as it should be. When ethical issues and policy decisions are at stake, sound principles and solid arguments are what count.

Yet formal analyses, as I mentioned earlier, leave out the personal factor. I have always been frustrated at not being able to tell about the dramas I most often have witnessed and much less often played a small role in. I attempt to redress the balance in this book by presenting the cases in dramatic detail, while also addressing the central moral issues.

I believe that something else emerges from a detailed, humanized, and dramatized presentation of the cases. These cases have taught me about people and the way they act when confronted with mortality. What I have learned is that people most often seem to be at their best when facing their worst fears. They may be puzzled by moral questions, but they rarely lack moral courage.

I think that anyone who reads these cases will also arrive at this conclusion.

The Woman Who Decided to Die

ONE

The Woman Who Decided to Die

S HE HAD ONE of the best rooms in the hospital. It was a single on the sixteenth floor that overlooked the long, stretched oval of the pond in the park below. Light from the sun's low-afternoon angle turned the water into a shimmer of pearl and pink and brightened the greenness of the surrounding lawn.

Her name was Traci, which seemed a frivolous irony given the grave nature of her disease. She had a form of acute myeloid leukemia that had not responded well to the standard regimen of chemotherapy. Traci, in the peculiarly accusatory language of medicine, had "failed" the treatment. The description seemed to imply that chemotherapy was a test she could have passed if only she had worked harder.

Dr. Samuel Wontage was one of her physicians. He was a short man with quick, nervous gestures and a smile that flashed on and off and yet never lighted up his sad, chubby face. Sometimes the smile, as if controlled by a defective switch, flashed on for no apparent reason, then disappeared just as suddenly. I decided that the smile was only another of his various nervous tics, like his habit of rubbing the side of his nose with his finger.

I was not, however, in a good position to have any reliable opinions about Dr. Wontage, because I had met him barely twenty minutes earlier. I was at that time developing an academic specialty in medical ethics and was eager to gain some direct experience. My knowledge of hospitals, doctors, patients, diseases, and treatments was largely secondhand, acquired almost exclusively from books. I hadn't even spent a night in a hospital as a patient.

I explained my situation to a friend from college who had become a research scientist at a medical school near my university, and he put me in touch with Dr. Wontage. He was a clinical professor, and I went to see him at his medical school office. We chatted a few minutes about my background in biology and the philosophy of science, which he was interested in. I told him I was hoping to learn about medicine and the sorts of decisions doctors and patients have to make.

"I'm going to see a patient now," he said. He turned on one of his smiles. "You're welcome to come along. She's got a big decision to make, and you might learn something from her."

I followed Dr. Wontage into the elevator, and he briefed me on the way up to her floor. Traci Williams was thirty-one years old. She was a file clerk in a state land-deeds office and had only a high-school education. She was married to an auto-parts warehouse worker who was three years younger and had never finished high school. They lived in Cameron, a small town three hundred miles away, and were the parents of two young boys.

Tim, Traci's husband, had driven her to the hospital and back during the six months of her treatments. He had lost a significant amount of his pay, because he'd had to take so much time off from work. She had been hospitalized for the last three weeks on the suspicion that the chemotherapy drugs had damaged her kidneys. But the injuries had healed and her kidney function was almost normal.

Tim now visited her on Saturdays and Sundays, but the kids were only three and five, and the likelihood of their harboring germs that could threaten their mother's life meant that she couldn't see them. Tim had been forced to pay a babysitter to stay with them. His parents lived in another state, and hers had both died in their fifties, her mother from the same type of leukemia.

I listened to Dr. Wontage talk and didn't ask questions. I was nervous about meeting Traci, because I felt intrusive. I couldn't do her any good, so why was I there? Was my going to see her like James Boswell's visiting Bedlam to marvel at the lunatics? I hoped not. But could I count on learning anything that would, even in the future, do anyone any good? It didn't seem likely.

Dr. Wontage rapped on the closed door of 1615, then pushed it open without waiting for an invitation. I followed him inside. Traci was sitting in the cushioned armchair beside the bed and looking out the window. I noticed how spectacular her view was. Yet I thought the shimmering oval of the pond looked strange against the flat green of the grass, as if it were the special-effects entrance to another dimension in a low-budget movie.

As we came into the room, Traci shifted her gaze. Her face brightened, and she looked at us with open interest.

"This is Dr. Munson," Dr. Wontage said. He spoke in a brisk, curt fashion that I soon learned was common among physicians, nurses, and even appointment secretaries. "Dr. Munson wants to find out how we do things in medicine, and if it's all right with you, he's going to sit in on our talk."

She was pale, her skin almost translucent, but her eyes were dark and tawny, like polished mahogany. Her hair reached below her ears and was a light shade of auburn, but it was obviously not her own. Under the wig, I imagined, her scalp was a stubbled field of new growth, a testament to the ravages of chemotherapy. Her nose was too rounded and her chin too sharp for beauty, yet her dark eyes shone with intelligence and curiosity.

"I don't mind," she said.

She smiled, but the smile didn't light up her face any more than Dr. Wontage's transformed his. Her voice was strong, though, which surprised me, because she looked so frail. Over a blue hospital gown, she was wearing a pale pink bathrobe tied with a polished satin sash. Her wrists, jutting out of the wide satin-trimmed sleeves, were as thin as a child's, and her long fingers were bleached and waxy. Even the nail beds were colorless, as if her entire body had been drained of blood.

Dr. Wontage stood in front of her and wrapped a hand around one of her wrists. "If it's okay, I want to examine you briefly."

That doctors always touch patients was something else I soon learned. They say they want to examine them, but usually they learn little or nothing from the exam. They use the opportunity to perform the physician's ancient ritual of laying on hands, and it definitely does some good. People who are well avoid contact with the sick, even sick family members, so touching patients reassures them and makes them feel cared for.

"Do I need to get on the bed?" Traci asked.

"You're fine where you are."

Dr. Wontage felt her pulse, then took a stethoscope out of the side pocket of his long white lab coat. He slipped the diaphragm of the instrument under her hospital gown and placed it on her chest. He then moved it around to several places, his face quiet with concentration.

While he was making his examination, I studied the pictures on her bedside table. Enlarged snapshots of a toddler in red shorts and a somewhat older boy in a blue hockey jersey with yellow numbers were opposite each other in a hinged gold frame. The toddler looked very serious, but the older boy was grinning. Both boys had hair so blonde it seemed to glow.

So did the man whose picture was in a small acrylic frame slightly behind the metal one. He was good looking in a clean-cut, boyish way, and his expression was as serious as the toddler's; maybe it was a worried look. Given Traci's age, I calculated that her husband was twenty-eight, but in the photograph he looked like a teenager. He could have been the older brother of his children.

"How have you been feeling?" Dr. Wontage asked Traci, still listening to her chest.

"All right, except for being tired," she said. "Sometimes I feel completely well and wonder what I'm doing in the hospital."

"I'm sure you do," Dr. Wontage said. He smiled, then folded his stethoscope and slipped it back into his coat pocket. Dr. Wontage sat on the edge of the bed. He had to give a little hop to get up that high, and his feet dangled an inch or two above the floor. I propped myself against the windowsill, half standing, half sitting, and crossed my arms over my chest, prepared to listen.

"I've been looking though your pathology reports," Dr. Wontage said. He rubbed the side of his nose. "You remember we talked about how we wanted the chemo to force your disease into remission? And the way we'd know that happened was if the number of blast cells in your bone marrow became normal?"

"Yes." She drew out the word, making her answer tentative, as if suspecting a trick. Her hands were laced together in her lap, her fingers locked, and keeping her shoulders rigid, she seemed braced for a blow.

"You did have a response," Dr. Wontage said. Traci's grip loosened. "Your blast cells are down, but they aren't where we'd like them to be."

"What does that mean?" She was startled by the news, and her fingers tightened again. She cleared her throat with a harsh cough. "Are you saying I'm not in remission?"

"I'm afraid you're not. Almost, but not really." Dr. Wontage was no longer brisk. He was reluctant to say what he felt he must, and he tapped his fingertips together, as if performing some small ritual for seeking forgiveness.

"So I still have cancer," Traci said. It was not a question. Her shoulders slumped, and she turned her head and looked out the window at the shimmering pond.

"If you were in remission, we would recommend a bone marrow transplant," he said. "It's a demanding procedure that requires destroying a patient's bone marrow with chemo and radiation, then repopulating it with normal tissue-matched cells. The procedure itself has a mortality rate of almost eight percent." He frowned. It's a bear, but young people do pretty well."

"Does it cure them?" Traci was skeptical, too intelligent to fool herself with hope. Her eyes bored into Dr. Wontage's, searching for the truth inside him.

"Not always," he admitted. "We're getting long-term survival of thirty to forty percent. The statistics are improving every year. Young children still do best, but adults under forty are moving ahead rapidly."

He made it sound like a race. Children versus young adults; young adults versus old people. Something about this way of talking bothered me. Then I saw that what he was saying applied only to categories, not to individuals. A particular woman with leukemia might or might not be cured by a bone marrow transplant. But she would have only one chance at it. If it failed, she wouldn't be alive to try again, so she would be denied the opportunity to "catch up" with the children who were doing better.

"But isn't bone marrow for people in remission?" Traci was not losing sight of her status. "That's not me."

"No, it's not," Dr. Wontage agreed. He crossed his ankles to stop his feet from swinging. "But we think a bone-marrow transplant is the best we have to offer you."

"Not more chemotherapy?" She seemed surprised; or maybe she was disappointed. I couldn't tell.

"It wouldn't pay off," Dr. Wontage said. He went back to tapping his fingertips together. "Once a patient has had two courses of chemotherapy and a bone marrow sample shows continued abnormalities, we can't expect a better outcome with more of the same." His lips twitched without forming a smile. "But a bone marrow transplant could make a difference."

"How much of a difference?" Traci sat up in her chair. She was alert and ready to take in every word.

"It's impossible to say exactly." Dr. Wontage rubbed the side of his nose, then wiped his fingers over his lips. "Maybe a bit less than for patients in full remission."

"Twenty percent?" Traci asked. She looked into Dr. Wontage's face, her eyes fixed on his again. "Fifteen percent?"

"Those are good guesses," Dr. Wontage said. He nodded, and his automatic smile flashed for a moment.

"Which one?" Traci seemed to be bargaining with him. She no longer held herself tight, but she wasn't relaxed. She was attentive, focused. "Higher or lower?"

"Let's say somewhere in between," Dr. Wontage said. He held up both hands as if surrendering to her. "That's a guess. A BMT isn't something we usually offer patients with your blast numbers." He hesitated and tapped his cheek with a finger before going on. "But you're so young. And you've got kids..." He trailed off, then shrugged.

Traci sat quietly, looking at Dr. Wontage. Tears trembled in her eyes, then trickled down her cheeks in small, silvery rivulets. She wiped them away with her index fingers. She pulled a tissue from the box behind the pictures. After dabbing at her eyes, she gently blew her nose with the same tissue. She wadded it up in her hand.

"What if I don't have any treatment?" Her voice was thick, and she sniffed without trying to muffle the noise.

"Patients with your disease, without treatment, rarely survive." Dr. Wontage spoke gently but resolutely. He wanted to leave no doubt in her mind about her prospects.

"How long do they live?" Traci asked. She was trying to sound objective and businesslike. She cleared her throat and dabbed at her eyes again with the wadded-up tissue.

"It depends on the person, like it does with the transplant," Dr. Wontage said. He touched his hair, brushing it back. "We can't be sure."

As he looked into her face he seemed to realize he was evading her question. His tone grew firmer as he finally said, "Maybe three to six months. Less if you...if someone with your disease, developed a serious infection. Chances are, we couldn't bring it under control."

"Thank you," Traci said. Her voice was softer again, smaller than ever. She lowered her eyes.

"We'll have to find a donor for you," Dr. Wontage said. Now he sounded upbeat. "Somebody whose bone marrow matches yours. Maybe your husband or even one of your children. A smile appeared for a moment. "We don't need much of it."

"I don't want the treatment," Traci said. Unwilling to catch Dr. Wontage's eyes, she turned her head away and looked out the window toward the glowing oval of the pond again.

"Talk it over with your husband," Dr. Wontage said, speaking gently. "I can meet with him and explain the way matters stand." He rubbed his nose. "You don't need to make a decision this instant, but we'll need to get started soon."

"I've decided," Traci said. Her voice was strong, and she turned to face Dr. Wontage once more. "I've already told you."

"Your insurance will cover the treatment," Dr. Wontage said. "You don't have to worry about that."

"Tim is young," Traci said. She spoke abstractly, as if talking to herself. "And he's not real good about taking care of himself. Not to mention the kids." Her eyes strayed to the photographs. "The sooner he marries again, the better it is for the whole bunch. The trips up here from Cameron, the worry and confusion...it's too much."

She bit at her lip, then swallowed hard. "I don't want to drag this out. Make everybody suffer, because I got sick. They've put up with enough already." She shook her head. "They need to get over me and get on with their lives."

"You have to talk this over with Tim." Dr. Wontage's voice was encouraging, almost insistent.

"No, you're wrong," Traci said. "What could he say, except go ahead with the treatment and do everything possible?" She dropped the wadded tissue into the wastebasket beside her chair. "I don't believe in that. If your time has come, it doesn't do anybody any good to drag it out." She raised her right hand as if taking an oath. Or maybe testifying to her beliefs.

"But you've got a real chance to make a long-term recovery," Dr. Wontage said. His face registered disbelief and alarm. The discussion was not going the way he had expected. Patients were supposed to accept the treatment their doctor recommended. Asking for their consent was supposed to be merely a formality.

"I guess the chance is real." Traci gave a little snort. "Real like I could win the lottery is real."

"The odds are better," Dr. Wontage said.

"I'll take your word for it," Traci said. "But they're not good enough. And what if the bone-marrow thing works? It won't cure me, so I could get sick again in a month or a year. Or even say two years." She shook her head almost violently. "That's not enough time. That would make it even harder on the kids when I died. And harder on Tim too."

"Some people live for much longer," Dr. Wontage said. His tone was sympathetic now, no longer insistent.

"Maybe the ones in remission," Traci said. "But you can't tell me I could count on having the same chance they do."

"No," Dr. Wontage said. Now his fingers were laced together the way hers had been while she was waiting to hear what he would say about the pathology reports. "I can't guarantee it, but it's possible."

"Possible isn't good enough," Traci said. She glanced at me for the first time, then looked back at Dr. Wontage. "I hope you'll both excuse me, but I'm getting real tired and don't want to talk anymore."

Dr. Wontage said he would be back to see her in the morning. I told her goodbye, and she nodded to me. But her mind was already taking her thoughts in another direction.

"That caught me by surprise," Dr. Wontage told me in the elevator going down. "Somebody with a good husband and small children usually doesn't need persuading. Typically, it's the other way. They grasp at straws, and you have to get them to see that some extreme treatment isn't in their best interest."

"Do you think she'll change her mind?" I asked.

"Oh, I believe so," Dr. Wontage said. "As soon as she hugs those kids."

I thought about Traci off and on during the next few weeks. She was so young, younger than I was by five years. I mused about her husband and children, speculating about what they might be like based on nothing more substantial than her few words about them and the photographs on the nightstand: the two little boys with hair so blonde that it glowed; Tim who needed somebody to take care of him as well as the kids; her impulse to get her dying over with to make it easier on her family; her willingness to become the dimmest of memories for her children to spare them a lifelong sense of loss. *I can barely remember my mother*, I imagined the older boy telling his own children. *She died when I was five.*

I was introduced to Traci in January, and I didn't see Dr. Wontage again until the following spring, more than four months later. I ran into him coming out of the hospital library. He didn't recognize me until I reminded him of our earlier meeting. While we were chatting in the hall, I asked him about Traci. Had she changed her mind about the bone marrow transplant?

"She didn't want to know any more about it than I told her the afternoon you visited," he said. He pursed his lips and shook his head. "And she ordered me not to discuss it with her husband. She said it was her decision, and she wasn't going to foist it off on him."

"So how is she doing?"

"She died," he said. "In late March, if I recall." He wiped a finger across his lips. "She was admitted to the hospital in Cameron with a lung infection. She refused to be put on the ventilator, even when her lungs were failing and she wasn't responding to antibiotic therapy."

So once again, I thought, Traci had "failed" treatment. It was medicine that had failed her, of course, but in the end she had achieved a sort of quiet triumph over it. She had died on her own terms, not on those dictated by her doctors and their treatment protocols.

Even though more than twenty years have passed, when I glimpse the sun glittering on the surface of the elongated oval of the pond in front of the hospital, I think about Traci. Her boys must have grown into men, and Tim must be middle aged. I feel sure that he remarried, but I wonder if he or his sons know what Traci did for them. I'm sure she never told them, because that would have defeated her purpose.

The basic rule of medical ethics is disarmingly simple: People should be allowed to make their own decisions. This is what philosophers call the *principle of autonomy*. If "do no harm" is the Prime Directive of medical practice, "get the patient's consent" holds the same position in medical ethics. Except in emergencies, a doctor can't so much as put a Band-Aid on a cut finger without the patient's consent.

Traci Williams's refusal to consent to further treatment surprised and dismayed Dr. Wontage, but he was right to respect her decision. We act autonomously when our actions result from our decisions. Thus, our autonomy is violated when our behavior is in any way coerced or manipulated. Traci's decision was her own, made freely and with knowledge of the relevant facts. She acted autonomously.

Our society is committed to recognizing the autonomy of individuals. We let people choose how they want to live their lives, even if we think their choices are foolish. We may sneer at the man who devotes his time and money to collecting Boston bus transfers, but we don't force him to give up this pursuit and do something more useful (like feeding the homeless) or more conventional (like working at Starbucks). We simply shrug and say, "It's his life."

We place restrictions on autonomy only for compelling reasons. Most of these reasons concern protecting others. We thus don't condone muggings, assaults, or sexual exploitation. John Stuart Mill argued that preventing harm to others (the "harm principle") should be the only grounds for limiting individual liberty.

Most social philosophers, however, believe it's necessary to go beyond this in order to promote and protect the interest of the society in which we live. Hence, we require people to pay school taxes (even when they don't have children), serve on juries, license their cars, and register for the draft.

Our society endorses this more expansive view of what constitutes a legitimate restriction on individual liberty. So we're always debating whether some proposed or existing restriction, such as laws against using medical marijuana or forbidding restaurants to cook with trans fats, can be justified. The question is always: "Is the state interest here sufficiently compelling to justify violating the autonomy of individuals?"

What is usually not disputed in our society is that it's wrong to force competent adults to act in ways that others (including the state) consider to be in their own best interest. We take a hands-off approach when it comes to self-regarding behavior—that is, behavior that concerns only the individual.

Hence, we reject the idea that someone who doesn't get enough aerobic exercise should be compelled by the government to work out in the gym "for her own good." Enforcing such a requirement would be a violation of the individual's autonomy, even if the motive behind it were benign. (This is why some people rail against corporations that pressure their employees to get into better physical condition by penalizing them if they don't.) Such a government requirement would be paternalistic, a manifestation of the "father knows best" attitude that once ruled in the traditional doctor-patient relationship.

Because Traci refused to consent to additional medical treatment, it would have been wrong for Dr. Wontage to force her to submit to it. It was his duty to lay out the medical options for her in a fair, complete, and informative fashion, to make sure she understood them, and to answer her questions as best as he could. He then had to leave the choice of an option up to Traci. It would have been wrong for Dr. Wontage to downplay the seriousness of Traci's medical situation, trick her into agreeing to another round of chemo, or lead her to believe that more chemo was very likely to drive her disease into remission. It was his duty not to decide for her, but to put her in the position to make her own free and informed decision.

Dr. Wontage played it straight and avoided paternalism. Even if he expected and wanted Traci to give chemo another try, he didn't impose his opinion on her. The life at stake was Traci's, and the decision about additional treatment was hers to make.

Not everyone in the same circumstances would have decided as Traci did, but that's the point of being free to exercise autonomy. By our decisions, we shape our lives, even, sometimes, to the point of refusing medical treatment and letting a disease take its deadly course.

Traci had her reasons.

TWO

Like Leaving a Note

ERE'S THE BEGINNING of the story as I later heard it:
Susan Winters, at twilight on an overcast November day,
was crossing Midway Boulevard, walking home from the
branch library where she worked. Either she didn't see the black van
speeding toward her or she believed it would stop for the red light.

Whatever the reason, she was caught directly in its path.

The front edge of the bumper struck her at hip level, tossing her
into the air like a floppy doll. When she fell, the back of her head struck
the concrete curb of the center divider. She lay sprawled on the street,
unconscious and bleeding from her injuries. A cashier in a sandwich
shop who had seen the accident called 911, and within twelve minutes
an ambulance arrived.

The paramedics made sure that Susan's airway was clear and gave her
nasal oxygen. They checked her blood pressure, injected a drug to boost
it, then started a saline IV. They put a cervical collar around her neck to
protect the upper part of her spine, then slid her onto a backboard. With
their siren bleating, they rushed her to the ER at Midwestern Hospital.

I was watching when the EMS crew backed their vehicle into a slot
at the arrivals dock, unloaded Susan's stretcher, then wheeled her into
Trauma Room One. This was several years before TV shows featured
emergency-room dramas that made the names of drugs and medical procedures as familiar as the names of the actors in leading roles.
Everything that happened that night was new to me, and I studied it
with wide-eyed wonder and an often quickened pulse.

That I was at the scene was more or less an accident. My friend
Charles Ying was a Fellow in the relatively new specialty of emergency

medicine, and he had invited me to spend an evening at the hospital where he worked. Charles was short and wiry, laconic, and intensely intellectual. He had been an undergraduate at Columbia, where I had met him, and he had gone on to get his medical degree from Harvard.

Charles was sometimes defensive about his specialty, because it had none of the intellectual cachet of traditional fields like internal medicine, neurology, or cardiology. Academic medical centers in the late 1970s tended to regard emergency medicine as differing little from a practical craft like auto mechanics. Each took skill and experience to perform well, yet, traditionalists intimated, neither required an understanding of underlying scientific principles.

"Just come and observe," Charles said. "You'll see how we save lives every day of the week, and that's not something the snobs can say. Most don't save one a year, if that."

I took up Charles's invitation more to please him than to satisfy my curiosity. My background was in biology and I specialized in the philosophy of science, but I knew little about medicine, nor was I particularly interested in it. All that was to change soon, but I didn't realize it at the time.

I arrived at Midwestern Hospital at six in the evening. Charles had me sign a form promising to protect the confidentiality of any patients I might encounter. He then printed my name on a plastic ID tag that had *Professional Visitor* in red letters across the top. I hung the chain with the tag around my neck. I felt a bit foolish, fearing that somebody might mistake me for a physician and ask me a medical question or, even worse, expect me to do something to help.

"This will let you go anyplace," Charles said. He smiled. "But I know you won't get underfoot."

To my surprise, Charles then apologized and departed to attend a Fellows meeting, leaving me on my own. I didn't have anybody to interpret for me, and only later was I able to make sense of much of what I saw during the evening.

I was so afraid of getting in the way that for a long while I stood against the wall behind the admitting desk. I watched the four women clerks as they briskly and efficiently gathered information from patients, then passed them on to the triage nurse. She asked a few more questions, then told them to take a seat in the crowded waiting area. Aside from the fact that so many people wore expressions of obvious pain or had bandages wrapped around their limbs or heads, the area could have been a departure lounge at an airport.

The majority of people who came to the admitting desk were sniffling or wheezing or complained of cramps or dizziness. I was surprised by the number of problems that seemed trivial, although I realized that I lacked the expertise to make such judgments. Maybe the middle-aged man who said he felt so lightheaded that he couldn't watch TV was only anxious. But then maybe he had a brain tumor. I began to understand why doctors order so many tests.

A highway-construction worker walked in with a gash in his chin caused by a chip of flying concrete. He was holding a red bandanna against his chin, but the front of his yellow T-shirt was splotched with blood. He was taken to the back and put into one of the examination cubicles to get stitches.

A frail, elderly woman with chest pains was also seen immediately. Her face was contorted and her eyes were narrowed, and she had a pallor that even her rouge and powder couldn't hide. She was helped up to the desk by her overweight daughter, who did most of the talking. A young hospital aide in a blue jacket brought in a wheelchair, and the older woman, accompanied by her daughter, was transported to a cubicle, where she was examined by a resident.

Like the triage nurse, I considered both those cases genuine emergencies. I followed the construction worker to the cubicle and stood just inside the curtain. I watched the resident, a slender Asian woman, examine the man, check his vital signs, and take his medical history. She asked him to remove the wadded-up bandanna he was pressing against the wound. The two-inch gash was ragged at the edges and dark in the center. It was still oozing blood. The resident tore the wrapper from a gauze pad and wiped off the skin surrounding the gash with brisk, deft strokes. She then injected some numbing medicine and began to sew together the edges of the torn skin with a small curved needle. She clipped off the ends of the stitches as she tied each knot.

"Trauma team to One," I heard a nurse say. She was speaking to someone in the next cubicle. "ETA three minutes. Female struck by car. Head injury and unconscious."

This sort of catastrophe, I thought, has got to be what emergency medicine is set up to cope with. Charles wants me to see how it responds to such a calamity, not how it handles people with sniffles or even gashed chins. Leaving the nimble-fingered resident to finish her stitching, I walked back through the treatment area and stood near the covered arrivals dock.

The boxy ambulance, its red and blue strobes flashing with blinding brilliance, turned into the hospital driveway. The driver expertly maneuvered the vehicle into the bay, and the paramedic riding in the back threw open the rear doors.

The driver joined the other paramedic to help unload the stretcher. They raised the stretcher to waist height, locked the wheeled frame into place, and rolled it into the hospital. The bag of clear IV fluid hanging from a rod attached to the metal frame swung like a pendulum. A tall black woman in green scrubs guided the paramedics through the broad doors of the trauma room.

The patient was Susan Winters.

She lay motionless on the stretcher, her torso and legs covered by a stiff gray blanket. Her skin was pasty white, with a faint blush of blue along her cheekbones. A tangle of dark hair hid her forehead, and a clear plastic oxygen mask held on by elastic straps obscured the center of her face. Her eyes were shut, and her cheeks were streaked with blood. The white cervical collar isolated her head, making it seem strangely detached from her body.

I followed the paramedics into the small trauma room and stood in the front corner beside the door. I had a close, unobstructed view of everything happening, yet I was also out of everybody's way. This was the first of what would turn out to be many fly-on-the-wall medical experiences for me. My muscles were tight, and my heart thumped rapidly. I felt vaguely guilty for being in the room as a spectator, a gawker at an emergency, unable to help, yet captivated by the drama.

The EMT crew pulled the stretcher parallel to a narrow examination table, and four of the eight or so people in the team slid their hands under Susan's body and shifted her from the backboard to the table. She was wearing black trousers and a blood-soaked beige sweater, but rather than trying to remove the clothes from Susan's inert body, a nurse used large shears to cut them down the sides, then pulled pull them away. A folded pale green sheet was laid across Susan's torso to protect as much of her privacy as practical. The skin on both legs was scraped and bloody, and her left leg jutted out at a strange angle.

"BP one-zero-five over sixty-five," the lead paramedic said, speaking quickly. "Pulse sixty-five, respiration eighty and shallow. Multiple injuries due to being hit by van and subsequent fall into street. Severe head trauma and unconscious at the scene. GCS one-one-four."

I learned later that GCS stood for Glasgow Coma Scale and the numbers represented a patient's responses. Susan's two "ones" meant that she hadn't opened her eyes and hadn't been able to say anything. The four meant that she drew back her arm from a painful stimulus, a jab with a needle or a sharp pinch. The sum of the three numbers was equivalent to a rough prediction of whether the patient was likely to recover. A total below eight wasn't promising, but I didn't know that at the time.

While the paramedic was talking, the trauma team moved into operation like a well-rehearsed dance troupe. Dr. Tina Evans (I learned names afterward) gave Susan a rapid examination, scanning every surface and looking into her mouth and ears. A nurse took Susan's aural temperature and read out the number when the thermometer beeped, but I didn't hear her.

"She's got clear fluid coming from her right ear," Dr. Evans said. Then she added, "It may be CSF," the abbreviation for cerebrospinal fluid.

A short, husky man in his late twenties pushed in beside Dr. Evans. He was Dr. Thomas Eagle, a resident in emergency medicine. He moved his stethoscope from place to place over Susan's chest, listening to her heart and lungs, then over her abdomen to listen for bowel sounds. He raised each of Susan's eyelids and checked her pupils as they were exposed to light.

"Reactive," he said. He put his head close to her face so he could use an ophthalmoscope to look through the pupils and examine her retinas. "We're getting some brain swelling," he announced. Swelling can cause the optic nerve to bulge.

Dr. Evans removed the plastic collar and used her fingers to explore Susan's neck without moving her. "We need cervical pictures," she said. "But I don't think the spine is displaced." She reattached the collar, then ran her hands over Susan's body, pressing on her abdomen and feeling the edges of her ribs and collarbone.

Dr. Morris, an older man in a long, white lab coat who had been standing to the side, approached the examining table. "Let's turn her," he said. Four people once again put their hands under Susan's body. They gently rolled her over, keeping her neck stable, then covered her again with the sheet. Dr. Morris parted Susan's hair at the back of her head and shined a penlight on the wound. Dusky white, red-tinged brain tissue oozed from between fragments of shattered bone.

I had never seen such a sight before, and I found myself holding my breath. I had smelled the blood from the first, but now its sickly sweet odor seemed to intensify. I felt myself waver and braced my back against the wall.

"Depressed occipital fracture," Dr. Morris said. "Call and say we're coming up for an MRI as soon as we get her stabilized. We also need to get a CSF sample."

A nurse in a flowered top used the red phone on the wall behind the table to relay the order. With Dr. Evans helping, Dr. Eagle bent Susan's body at the waist, bowing her back, then inserted a long needle between two lumbar vertebrae and into her spinal canal.

"Blood in the CSF," Dr. Eagle said.

"Not a surprise," Dr. Morris said. "Her skull fracture is as bad as any I've seen."

"Her left arm is broken," Dr. Evans said. "I also think her left hip is fractured and displaced."

"Get X-rays after the MRI," Dr. Morris said. "We also need to look for abdominal injuries. It's possible the hip absorbed the worst of the impact, but let's make sure her spleen isn't ruptured. Turn her back over."

The process was repeated. A nurse readjusted the oxygen mask, which had become dislodged during the maneuvers. Susan's face, despite the oxygen, still had a tinge of blue.

"Her blood pressure is unsteady," Dr. Eagle said. "And she's shocky. Let's get the whole blood started."

"I'm on it now," a nurse said. She was already hanging a plastic IV bag from the pole.

"Check the blood gases," Dr. Eagle said. He was obviously the person in charge, although Dr. Morris seemed to be monitoring his performance.

Susan lay immobile, the only fixed point in the swirl of motion around her. Then that changed in an instant. Her chest heaved, her hand came off the table, she sat halfway up, and her mouth gaped wide, knocking loose the oxygen mask.

"Get an ET tube in," Dr. Morris said. He was flatly calm and unflustered. He stepped back from the table to give Dr. Evans room to work.

Dr. Evans accepted the thin plastic tube a nurse held out for her. Another nurse cut the elastic straps securing the plastic oxygen mask and pulled it free. Dr. Eagle bent Susan's head backward as much as the cervical collar allowed and pulled down on her chin, opening up her airway. Susan twisted on the table, but three nurses gripped her body and held her steady. Dr. Evans, clutching a flashlight in her left hand, peered at the back of Susan's throat. She snaked the tube into Susan's trachea at the first try, and then I became aware of the steady mechanical sound of the ventilator.

"Make sure it's not in her stomach," Dr. Morris said.

"I'm certain I got it," Dr. Evans said. Even so, she placed the diaphragm of her stethoscope on Susan's chest and listened for sounds of air in the lungs. She frowned with concentration for a moment, then she looked pleased, almost happy. "Right where it should be."

Susan lay calm again, the hunger for air satisfied. One of the nurses taped the plastic tube into place.

The trauma team kept working. Following orders from Dr. Eagle, they gave Susan electrolytes and additional drugs to raise her blood pressure and strengthen and stabilize her heartbeat. They slipped over her legs and hips, at some stage, a pair of dark green plastic MAST trousers that were then inflated. The idea was that squeezing the legs would reduce peripheral blood flow and so help keep the brain perfused with blood and oxygen.

Susan was young and in good physical condition, but she was gravely injured. The broken left hip and arm were serious, but what worried the team most was the skull fracture. As soon as her blood gases, heart rhythm, and blood pressure were stable, Dr. Evans and two nurses wheeled Susan into the transport elevator to take her to the MRI suite. The emergency was at an end, but Susan's problems weren't over.

I didn't follow Susan and her escorts, but I was present when Dr. Evans came back and reported the results of the MRI to Dr. Morris. Dr. Eagle had left to treat other patients, and I had exchanged a few words with Dr. Morris while the custodial staff cleaned up the trauma room.

"She has massive damage to her occipital lobe," Dr. Evans said. This was due to the impact of Susan's head on the concrete curb. "Her meningeal arteries are torn, and she's bleeding into the brain spaces. Blood in the cerebrospinal fluid is confirmed, and pressure on the brain stem is likely to occur in the near future."

I had a feeling of dread as I listened, even though I had never met Susan and knew her only as an unconscious figure on a stretcher. The experience was like watching a movie in which you know a killer is hiding in a closet that someone is about to open. You can do nothing but watch with increasing horror.

I left the emergency area without knowing what Susan's doctors planned to do help her. When I met up with Charles later that evening, I sat with him in the hospital coffee shop while he ate a quick supper. I told him about watching the trauma team at work and said I was interested in following Susan's case.

"I'll find out what's happening," Charles said. He used a wall phone reserved for physicians to call Amy Wilson, a resident he knew who worked with Dr. Lawrence Backer in the neurosurgical intensive care unit. Dr. Backer had become Susan's physician when she was transferred from the ER. Charles' conversation with Amy Wilson was brief and one-sided.

"Susan Winters went directly from MRI to the operating room," Charles told me. "By the time she was prepped, the neurosurgical team was ready to roll. They tied off the bleeders and removed the bone debris and damaged brain tissue. They put her on IV drugs to lower her intracranial pressure and screwed a gauge into her skull so they could monitor pressure changes."

"So she's out of danger now?" I asked.

"I wouldn't put it that way," Charles said. "She's in the NICU, and she's stable. She'll be monitored closely, but a good outcome isn't guaranteed."

Feeling tired and emotionally drained, I said goodbye to Charles and left the hospital late that evening. Although I now accepted Charles's view of the importance of emergency medicine, I realized that it must exact a steep price from those working in the area. While falling asleep that night, I was haunted by images of Susan's damaged, bloody body.

I had no experience at the time of the cautious ways in which physicians phrase their judgments about a patient's chances of recovering from a serious disease or injury. Thus, I was surprised almost to the point of being shocked when Charles called me the next morning to tell me Susan was dead.

"The neurosurgeons were never able to get the brain swelling under control," Charles said. "Eventually, her brain stem was crushed by the pressure."

"And that's what killed her?"

"That was it," Charles said. "Dr. Backer and his residents ran through the standard protocol this morning. They checked her cough reflex, pupillary response, and pain response, and did a bunch of other things, including getting an EEG, then concluded that her brain had no organized electrical activity. They took her off the ventilator for five minutes, and when she didn't start breathing spontaneously and her blood pressure dropped, they put her back on. Her heartbeat was still strong, though, and her blood gases showed that her organs were getting sufficient oxygen."

Charles paused, apparently to give me a chance to ask questions, but I understood so little of what he was telling me that I hardly knew how to frame one.

"I got all this from Dr. Backer when a resident called him down to the ER to consult on a patient," Charles went on. "He's informed Susan's parents, and they've already said they don't want her to be kept on life support. He's giving them a chance to sit with her and say goodbye, and a nurse that works with the transplant coordinator's office is meeting with them at noon."

"Do her parents want to donate her organs?" I asked.

"That I don't know," Charles said. "But I told the nurse you were interested in the case, and she said you're welcome to sit in on the discussion, if the parents agree."

I hesitated. I had, naively, expected Susan to recover from her accident, having witnessed the intense and sophisticated efforts of the trauma team. Now she was dead. My connection with her was neither personal nor professional, yet I had an impulse to follow her case to the end. In a sense, that meant following it even past her death. This was perhaps the beginning of my interest in medical ethics, yet at the time, I couldn't say I had any professional end in view. I thus could have been accused, with some justice, of trying to satisfy my curiosity.

"Who's the nurse?" I asked.

"Lisa Samuels," Charles said. "She's near the transplant suite on the lower level."

I met Lisa Samuels in her office. She was short and pencil-thin, with a pixie face framed by black hair clipped short. When I introduced myself, she rose from her desk, then shook hands with me and smiled. Her smile was warm, but it didn't keep her from looking exhausted.

"I've already talked to the parents, and they're okay with you sitting in," she said. "They're with Susan now. I told them they should say goodbye, but not to stay too long. That would only make it harder on them." She shook her head. "People can get used to almost anything."

"Did they agree to let her organs be used?" I wasn't sure how to phrase the question. I'd heard Charles talk about "harvesting" organs, but the word seemed disrespectful, and I didn't feel comfortable using it.

"They haven't said no," Lisa said. She sounded upbeat but serious. She glanced at her watch, which she wore on her right wrist. "I said I'd meet them in the NICU lounge at noon."

"Can you tell me anything about them?"

Lisa picked up the clipboard from her desk. "The mom's name is Joyce and the dad is Dennis. Let's go up and see them, and I'll fill you in while we're walking."

The NICU lounge was at the opposite end of the hall from the nursing desk. Curved walls topped with glass blocks carved out the space, and the dark gray carpet that covered the floor and the first three feet of the walls produced a hushed atmosphere. The entrance was wide and had no door. Inside, you felt isolated from the rest of the world, but not secure. It seemed to be a place to be worried and sad.

Susan's parents were already waiting. Joyce was sitting on the near end of a long sofa, and Dennis sat in a chair by the far end. Joyce was gazing blankly at the opposite wall, and Dennis was staring straight ahead at the entrance. If lines had been drawn to represent where their eyes were focused, they would have intersected at right angles. No one else was in the lounge.

Lisa introduced me. I shook hands with Dennis and nodded to Joyce. I managed to say, "I'm so sorry about your loss." Joyce gave me a faint, sad smile, but Dennis showed no response.

"Let's sit together so we don't have to shout," Lisa said. She pulled over a chair to where Joyce was sitting. "Dennis, please come over and join your wife on the couch."

Dennis did as he was asked, and Joyce reached over and gently rubbed his right hand. He was, I knew from Lisa's briefing, the manager of a supermarket. He was in his late forties, with thinning brown hair combed straight back from a high, lined forehead. His aviator glasses had silver frames, and a scar the size of a dime made a shiny patch at the left corner of his upper lip.

I dragged up another chair, but I was careful to sit behind and to the left of Lisa. I appreciated Dennis and Joyce letting me be present, but I had no right to be part of the discussion.

"I want to thank both of you for talking to me," Lisa began. She was business-like but sincere. "I realize this is a hard time for you, and about the last thing you want is to have to make an important decision."

"I have trouble thinking of Susan as dead," Joyce Winters said. She squeezed her eyes shut, then put both hands over her face. "I know her brain is gone, but I halfway expect her to wake up and smile at me." She spoke the last words in a low, sobbing voice.

Joyce was stout, with plump arms, a smooth, round face, and frizzy blonde hair. You could see that when she as young she had been pretty. She now worked part-time, Lisa had said, selling advertising space in a local telephone directory. She was most likely a responsible, mature person, but now she looked like a little girl who was terribly upset by something she couldn't understand.

"If she's dead, why don't they turn off the machines?" Dennis said. He spoke in a flat voice, without anger and without a sign of any feelings at all, but he fixed his eyes on Lisa.

"That will happen when you tell Susan's doctor you want them to," Lisa said. "But at the moment, you have the opportunity to do something for other people." She kept her eyes on Dennis, then glanced at Joyce. Joyce was wiping her nose with a tissue, but she nodded.

"Susan wanted that," Joyce said. "That's why she signed the back of her driver's license. She wanted to donate her eyes and her heart and everything else. Whatever the doctors can use to help somebody else."

"That's what we're here to talk about," Lisa said. "If you already know your daughter's wishes, that might make the decision easier for both of you."

"I don't want her cut up," Dennis said. His voice was thin and strained. He rubbed the scar on his lip with his thumb, as if it hurt and needed soothing. "She's already been through enough, with her broken bones and her fractured skull."

"She signed her donor statement," Lisa gently reminded him. "That means she wanted to become a donor."

"Then why involve us?" Dennis said. His voice rose, taking on a sharp, angry edge. "She signed the license, so the hospital can do whatever they want without saying damn-all to us, can't they?"

"Legally, we can remove her organs," Lisa admitted. She spoke quietly to avoid antagonizing Dennis. "But we always ask for the consent of the immediate family. Always." She paused, keeping her eyes on Dennis. "Do you mind if I take a minute to tell you what the procedure is like?"

"That's why we're here, I guess," Dennis said. The anger had gone from his voice. He took a deep breath and sighed. "Go ahead." Then after a moment's hesitation, he added, "Please."

"People often don't know what their family member wanted to do about donating their organs," Lisa said. "Then it's hard for the family to decide, because they get involved in speculating about what the deceased person might have wanted."

"Well, we know already," Joyce said. She spoke so bluntly that she left no doubt about her own opinion.

"Even so, parents have a particularly hard time deciding," Lisa said. "Maybe it will help you if you known that Susan's organs could be used to extend the lives of four or five people."

She paused to give Joyce and Dennis a chance to absorb what she had said. "Many people are waiting for a heart or a liver or a kidney," she continued. "I'm talking about *hundreds* of people, thousands really. All of them are sick, and some may even die before they get a donated organ."

"In this hospital?" Dennis asked. He took off his glasses and polished the lenses on his tie, but he kept his eyes on Lisa. His imagination seemed caught by the idea that thousands of people were waiting for an organ to save their life.

"Some are upstairs right here," Lisa said. "Others are at home, relying on medicines and dialysis to keep them alive while they wait for an organ to become available."

"But Susan could help them," Joyce said. She addressed the remark to Dennis, sounding almost proud.

"If you choose to donate her organs, they would be put to the best possible use," Lisa said. "But I want to emphasize that you don't have to donate them. You can say no, and we'll honor your decision. You don't even have to give us a reason."

"It's completely up to us?" Dennis asked. He seemed surprised.

"Absolutely," Lisa said.

"Are the heart and so on sold to the people who need them?" Dennis asked. "I wouldn't want them just to go to rich people."

"They won't," Lisa said firmly. "It's illegal to sell organs. Patients have to pay for their surgery and hospital stay, but not for the organs." She paused, then added, "Some families find that helping others also helps them a little." She looked at Dennis. "I can't tell you it ever makes up for the loss."

"It couldn't come close," Dennis said. He spoke softly, as if talking to himself. "Not close enough to notice."

"Susan tutored children after school," Joyce said. "She helped kids with reading." She turned her head down.

"And donated organs go to those who need them?" Dennis asked. His tone was challenging. "They won't be thrown in the trash, after we've been through all this?"

"Not if we can help it," Lisa said, shaking her head. "But we might discover that one or more of Susan's organs has been damaged and isn't suitable for transplant. The drugs given to her might have injured her liver, for example. Also, even if her organs are in good condition, they might not all be transplanted."

"Why?" Dennis asked. He sounded suspicious again. "If so many people are waiting?"

"We may not be able to find a suitable recipient," Lisa said. "The recipient must have a blood type compatible with Susan's, and you can't give an adult heart or liver to a small child, for instance. And for kidneys, we also have to get the tissues to match up."

"You couldn't find the right kind of patient among the hundreds of people?" Dennis asked.

"Most likely we can," Lisa said. "But sometimes an appropriate patient may be too far away. When an organ loses its blood supply, it needs to be transplanted soon or it becomes unusable. For example, a heart has to be transplanted within about four hours. So it would be pointless to send it from Boston to a patient in San Francisco."

"Prepackaged meat," Dennis said. Tears ran down his cheeks, and he wiped them away with a bent finger.

I was slow to grasp the allusion. Then I remembered that Dennis was a supermarket manager. He must be imagining his daughter's organs being sent to hospitals the way meat from processing plants is delivered to grocery stores. Even if Susan's organs could save dozens of lives, no parent could contemplate such an image without becoming upset.

"Susan's body would not be disfigured," Lisa said. She either failed to catch Dennis' comparison or thought it best to ignore it and move on. She spoke in a matter-of-fact tone, but not without warmth. "The surgeons use the same procedures they use in any other operation."

"So we can get her back to bury her?" Joyce asked. "We already know where her grave will be." She stopped abruptly, then sucked in her lower lip, unable to continue. Finally, in an uncertain voice, she managed to say, "We bought two cemetery lots for ourselves. We never thought about needing one for Susan."

"Parents don't expect their children to die before them," Lisa said. "That's only reasonable." She took a pack of tissues from the pocket of her white jacket and passed them to Joyce. Joyce pulled out a tissue and wiped her eyes. She then blew her nose.

"Have we already made the decision?" Dennis asked. He sounded more puzzled than challenging now, and he turned toward Joyce. "You're talking that way."

"Dearie, we've got to do what she wanted," Joyce said gently. She rested a hand on her husband's leg. She looked at him with soft eyes, and for a moment their gazes met.

"I don't see why," Dennis said, sounding puzzled. "It's just us, and we can do whatever we want. She's not here to discuss it." He hadn't once used his daughter's name, as if he couldn't bear associating it with death.

"That's what I've been saying," Joyce said. Her tone was gentle but firm. "We've got to speak for Susan, and she told us what she wanted."

"She never told me," Dennis said. He ran his hands over his hair, sweeping it back from his forehead. He rubbed at the scar once more. "We never talked about morbid things."

"I didn't discuss it with her either," Joyce said. "But when she signed the back of her driver's license, that was like leaving us a note." She gave Dennis a steady look. "She didn't have to talk to us, because she'd arranged for us to get her message."

Dennis pressed his palms together and held his index fingers to his lips as if he were praying. He sighed and closed his eyes. His face seemed to sag, and he looked old and sick.

"I should leave you two alone for a while to talk matters over," Lisa said. "If you have questions, please ask me. If I don't know the answers, I'll find somebody who does."

Lisa started to stand up but Dennis held up a hand. "Wait a second," he said. "We don't need to prolong this."

Lisa stayed in her chair. Everyone looked at Dennis. He fixed his gaze on Lisa.

"Take her heart and liver and whatever else you need," Dennis said. "To be honest, I'm not saying this because I want to help people. I guess it's good if that happens." Dennis stopped talking, unable to continue. He and pressed his lips together and closed his eyes.

The few seconds that passed seemed like minutes. Joyce leaned her head against Dennis's shoulder. She slid her left hand under his arm and held on to it.

"But that's not my reason," Dennis said at last. He was talking to Lisa, and he kept his eyes on her. "I want to do whatever she wanted. That's all I care about, really."

"That goes for me too," Joyce said. "I think we're ready to sign whatever we need to."

I left before the paperwork was finished, and later that afternoon I called Charles to let him know how things had gone.

"I'm glad to hear the family made the right decision," he said. "She was apparently a good donor."

"That sounds a bit cold," I said. I felt vaguely offended, because I had watched Susan's parents struggle with the donation question when they hadn't fully accepted her death. Also, just by knowing what Susan had gone through, I had established some sort of relationship with her. I thought of her as a real person, not as an anonymous donor.

"I don't mean it to be," Charles said. "But we can't save every patient, and everybody did what they could to help her." Then he said, almost apologetically, "Unfortunately, the magic doesn't always work." It was an expression I was to hear from time to time in the coming years.

I promised Charles I would come back to the hospital some evening soon and watch him go about his job. "We really do save lives," he said. "Most cases don't turn out like this one."

Charles was right that Susan's case had ended badly—for her, her parents, and everybody else who cared about her. Yet, despite Dennis's insistence that he wasn't concerned about helping other people, Susan's foresight and generosity most likely extended the lives of several other people. Because she had helped make sure that something good came out of her tragedy, her case didn't end as badly as it might have.

Susan Winters's parents knew that she wanted her organs donated. Signing the statement on her driver's license was, as her mother said, like leaving them a note.

Such expressions of intention most likely have an enforceable legal status. Yet, as Lisa Samuels explained, unless a dead person's next of kin also consent, transplant centers don't remove the organs. The hospital could seek a court order allowing them to carry out the person's wishes, but the result would be a public-relations disaster. The lead story on the nightly news might begin, "An unconscious patient at a Chicago hospital will have her organs removed over the objections of her parents."

Not only would this damage the reputation of the hospital, but more important, it would depict the transplant community as crass and predatory. If this attitude became widespread, organ donations would drop, and the moral legitimacy of organ transplantation itself might be called into question. One life may be saved, but thousands of others lost.

SHORTAGE

The current organ transplant system is built around altruism. People like Susan Winters express their willingness to have their organs used when they die, or people still alive donate a kidney or a segment of their liver, lung, or pancreas to someone in need. But the number of people whose lives could be saved by a transplant is huge.

Some 100,000 people are on the transplant waiting list at any given time in the United States, and about 10,000 of these will die while waiting. This is roughly equivalent to *three times* the number of those who died on September 11, 2001, at the World Trade Center.

The waiting list continues to grow at a rapid rate. Yet the need for organs is already greater than can be met by relying on deceased donors. About 15,000 brain-dead potential donors are available annually, and if the current average of 3.6 organs were recovered from each, this would amount to 54,000 organs—about half the number needed now. Dismal though this estimate is, the actual situation is worse. Barely more than 50 percent of those asked to donate the organs of a brain-dead relative consent. Thus, only about a quarter of the organs we need *right now* are available. Altruism isn't producing enough transplant organs. Can anything be done to save the lives of some of the thousands doomed to die?

PRESUMPTION

One idea is to "presume" informed consent and remove organs from the dead without asking their families. Or organs of the dead could be declared state property and conscripted. Critics allege that both proposals violate the "due process" clause of the Constitution, which prohibits the government from taking anything from citizens without a judicial review. Dead bodies aren't legally recognized as property, but because they are by custom under the control of the next of kin, conscripting them might violate due process.

EXCHANGES

Increasing the number of live donors would do the most to save lives. Although liver, lung, and pancreas segments can be contributed by live donors, the contribution of kidneys would save the most lives. Sixty percent of all organs transplanted are kidneys, and people needing a kidney fill half the waiting list. If only one of every three thousand people became a live donor, the kidney shortage would be solved.

We are far from reaching this level of altruistic behavior, but the number of live donors has gone up in recent years, thanks to kidney exchange programs. Here's how they work. Suppose my wife needs a kidney, and I'm willing to donate one, but our tissues don't match. Once she would have had to wait for a kidney from a deceased donor.

But now I can swap my kidney for one from another donor. His kidney is a match for my wife, and mine is a match for his daughter.

Matching can involve several people. When A and B can't swap directly, if A swaps with C, C can then swap with B. As many as five donors and recipients have been involved in such round-robin exchanges. Critics object that exchange programs treat organs as "commodities" by eliminating the requirement that donors have a special relationship (family member, close friend) with recipients. In my opinion, that hundreds of lives can be saved by kidney exchange programs reveals this as a feeble criticism.

MARKETS

Bartering one kidney for another constitutes a market. Thus, an obvious variation of the practice would be to allow people to exchange a kidney for money. But the 1984 National Organ Transplantation Act makes it illegal to sell organs. Hence, the law would have to be repealed or modified if we want to allow those in need of a kidney to pay a live donor for one his.

The open market can be cruel, however. If kidneys were offered on e-Bay, prices might be bid so high that only the rich could afford them. To avoid this, kidneys could be sold in a regulated market. An organization like, say, the Organ Distribution Network (ODN), operating under a federal contract, could buy kidneys at a set price, then sell them at a set price. This could put a kidney within the reach of people of modest means, and those unable to come up with the money could count on Medicaid to help them.

ALTRUISM AND DIGNITY

Allowing kidney sales isn't incompatible with altruism, because anyone wanting to donate a kidney without payment could. Even so, the possibility of selling a kidney may make unpaid donations less likely and so reduce the amount of altruism in the world. What seems likely, though, is that a kidney market would result in a net increase in the number of kidneys for transplant.

A kidney market unquestionably involves commodifying the human body. A volunteer sells her kidney in much the same way as she might sell her watch or her car. Isn't there something debasing about treating a kidney like a watch? Isn't this a blow to human dignity?

Perhaps. Yet focusing on altruism and dignity ignores the awkward fact that thousands of people are dying because they need a kidney. If the one out of every three thousand people required actually donated a kidney, this would meet the need and be a proud testament to our best ideas about ourselves. Unfortunately, we have little reason to believe this will ever happen. One way to respect human dignity is to save thousands of lives. If paying live donors can do this, I think the result trumps the value we place on keeping the human body out of the marketplace.

Would selling one's kidney be morally wrong? I fail to see how selling something that you're free to give away could be wrong. It would be wrong to trick someone into selling a kidney or to lie about the risks it involves But if the conditions of informed consent are satisfied, I can't see why it would be any more wrong to sell your kidney than to sell your car. Both transactions are cases of autonomous decision-making.

DECEASED DONORS

What about the organs of deceased donors? We could continue to depend on the altruism of their next of kin, but perhaps deceased-donor organs should also be available on the market.

Families of deceased donors often complain that everyone involved in organ transplantation makes money, except donor families. (In one famous case, a mother had to borrow the money to bury her daughter after she donated her daughter's organs for transplant.) Surgeons, nurses, coordinators, organ procurement organizations, and hospitals all make money from transplants, and only the families of donors are expected to behave altruistically. This could change if we permitted a market in transplant organs from deceased donors.

This could work by using the same regulated market employed in acquiring and distributing kidneys from live donors. ODN could buy transplant organs (heart, liver, lungs, pancreas, and kidneys) from the families of the dead at a fixed price, then sell them to patients at a fixed price. No competitive bidding; no deals between private citizens.

FLAW

The ODN plan can put donor organs within the financial reach of people who aren't rich, but it doesn't allow a live donor to make as much money as possible. ODN gives the donor a take-it-or-leave-it price, but she could benefit most by offering her kidney in a market in which competing buyers bid up the price.

This is a serious shortcoming of the ODN plan, considering that the person willing to sell a kidney isn't likely to be Donald Trump. That the poor, not the rich or middle class, are the ones likely to sell their kidneys is the most significant objection to an organ market.

But it's also the poor who sell their labor and work at jobs that are among the most dangerous, dirty, and difficult. We don't force people to work in coal mines, but we offer them the possibility. Ultimately, whether someone takes the job is a matter of exercising autonomy and choosing among available alternatives. The situation isn't so different when someone has to decide whether to sell a kidney.

Thousands are dying for lack of transplant organs. Thousands of usable organs are buried or cremated. Yet even if we could take all the useable kidneys from the dead, we wouldn't have enough. We need more organs from both deceased and live donors.

Altruism isn't a plan for success, but buying and selling organs is illegal. It's time to repeal the law and develop a plan that will strike a balance between exploiting the poor and restricting benefits to the rich. Thousands will die before we can accomplish this, but thousands of others will live once we are successful.

THREE

The Agents

NANCY TRAIL WOKE up feeling terrible.

The muscles in her neck and back were stiff, and her stomach felt bloated and unsettled. She had slept badly and had a dull, low-grade headache. Her body felt so heavy that getting out of bed and putting on her clothes involved a series of small struggles. Each task took all the will she could muster, and she constantly felt on the edge of failing.

She zipped up her jeans, then searched the closet for her long-sleeved white shirt with the button-down collar. She put on the shirt, then pulled on a thick blue sweater. She wore the shirt on bad days, because it seemed to help. Maybe because white was for purity and the collar buttons kept things in order.

She felt especially bad this morning, but she knew what was wrong. The Agents had injected her with gold while she slept. They had been doing it for days, maybe weeks. She had lost track of the last time she felt really well. Maybe as long ago as a month. She'd been all right at the beginning of the semester, but then the Agents had started working on her. Now her body felt heavy and clumsy because it was filled with gold.

She turned over her left wrist and held it close to her face. She could see gold in the veins running into her hand. The gold was denser, more solid, than ever, which was why she felt particularly terrible today. The thin skin of her wrist made it possible to see the gold in her veins, but it was also where she couldn't see it. The Agents injected the gold into her bloodstream, so it was spread throughout her body.

Gold was even circulating in her brain. That's what made it so hard to think. When she tried to focus on something, dozens of thoughts flooded her mind and clamored for attention. The commotion was so distracting that she couldn't think clearly. Her thoughts fragmented and fuzzed, making it impossible for her grasp anything. Sometimes she cried from frustration and confusion.

She wondered how her mother had coped. Once, when she was five or six, her mother had told her about getting injections of gold. For arthritis, her mother had said. She'd believed the story at the time, but now she realized it was really the Agents who had filled her mother with gold. Gold was powerful, and the Agents had used it to control her mother. Now they were doing the same to her.

Her first class was at ten o'clock, and it took her ten minutes to walk from Williams House to Dewey Hall. That left half an hour for breakfast and skimming the assigned reading. She would have to hurry, because she didn't dare be late.

The Agents hated her being late and would criticize her remorselessly for it. "You're irresponsible," they would say. "If you can't get to class on time, you don't deserve to be at an Ivy League college. You're a fraud and a disappointment to your mother." She knew they were right, and that's why she had to avoid their criticism. She couldn't bear listening to them spell out her weaknesses in painful detail.

Her suitemates were already gone, so she had the tiny kitchen to herself. She switched on the electric kettle, put a slice of bread in the toaster, and got out a mug and a tea bag. She still found it hard to move, and she spilled boiling water on the countertop as she filled the mug. She seemed to be living under water. The water slowed her movements and cut her off from the world, making everything around her remote and unreal.

She took her toast and tea to the dining table. While she ate, she looked over her Western Ideas assignment. She hadn't read the Mill selection yet. Ten pages from *On Liberty* wasn't much, so she could skim it before leaving for class. But she found concentrating impossible. Her mind slipped out of focus the way she hated. The printed words lost their meanings, and she no longer knew what she was reading. Then the Agents began to whisper to her. Because the gold was building up in her brain, they were speaking more often than even a week ago.

The voices came from the small TV at the end of the table. The TV wasn't turned on, but that didn't matter. The Agents controlled its speaker the way they controlled the stereo in her room. Sometimes when she was in bed, they would even speak to her through her pillow. They could reach her anytime they wanted.

The Agent speaking was the one called Ostra. The others were called Egon and Tuve. Egon was the scariest, with a deep, growling voice. Tuve seemed hesitant, almost shy, yet he could say hurtful things. Ostra always did most of the talking, though.

"You need more sex," Ostra said. His tone was reasonable but insistent. "It would make you happier. Why don't you screw Bill Hanley? He's cute. Wouldn't you like some of that, hmmm?"

She kept her eyes on the book. She always pretended not to hear, because she didn't want to encourage them. That was what her mother used to say about petting the dog when he jumped up on the sofa: ignore him, and he'll go away.

"Go do it now," Egon said. The voice was like a snarl, and it was filled with contempt. Contempt for her. "Go find Bill and tell him to do it to you."

"You're like a bitch in heat," Tuve said. The remark was followed by a sigh. Then with a giggle he added, "It's disgusting."

She tried to shut her mind. The Agents talked about sex so much and in such crude terms that it embarrassed her. She hardly ever thought about sex, and she certainly didn't discuss it with others. Why the Agents should mention Bill was a mystery, because she barely knew him. He was in her biology class, and once he'd borrowed her notes. He'd given them back at the next class and thanked her. That was the only conversation they'd ever had.

Even so, during the last couple of weeks, the Agents had kept on bringing up Bill Hanley. Sometimes what they said about Bill and her and what they ought to do together were so explicit and gross that it mortified her. She would feel herself blush.

What they said also made her self-conscious. She could no longer allow herself to look at Bill. Last week, when the class topic was sexual selection, she'd had to leave the room. She knew Bill could tell what she was thinking, and he was going to force her to have sex with him. If she'd stayed in the room, she'd have started screaming at him. She didn't want to be raped, even by somebody as good-looking as Bill.

She took a sip of tea, and it tasted bitter and metallic. Probably the gold. The Agents were still forcing it into her. They wanted total control.

Already they were forcing her think about sex all the time. Last week, when she'd gone by herself to the Brass Monkey for a beer, they'd made her go up to a stranger, a blonde guy with a ponytail, and proposition him.

"Do you want to have sex with me?" she'd asked right out. His eyes got large and he looked shocked, but before he could say anything, she ran out the door. She heard the people standing with him at the bar laughing. She knew they were laughing at her. The Agents told her what a fool she'd made of herself, how she couldn't do anything right.

The next time she might not be able to run away, and that scared her. The Agents might make her come on to another stranger, then force her to go through with it. She wasn't going to be able to resist them forever.

She left the dishes on the table and put on her long leather coat. The black coat, with its turned-up collar and covered buttons, usually made her feel stylish and confident, but now she buttoned it automatically. Her thoughts were elsewhere. She put her Western Ideas text in her shoulder bag and got her purse out of her room.

While standing in the entrance hall, she realized with certainty that she couldn't go to class. Class would be pointless. Her life was pointless. She'd been going maybe half the time for the last two weeks. Bonnie, Karen, and Jeff, her suitemates, had been worried about her. Karen and Bonnie had also been hurt by the way she had started avoiding them. They told her how much they missed the fun the three of them used to have together. That had made her sad, but it was to hard to be with people. It was easier, somehow, to be with the Agents.

She put her hand on the doorknob, then drew it back.

She dropped her book bag and purse on the floor, then slipped out of her leather coat and let it fall. She then crossed the living room and raised the window. She leaned out and looked down on the street. Far below was a miniature city scene, so perfect and natural it could have been a toy-store display.

A UPS truck was double-parked in front of the Dunkin' Donuts on the corner. People in jackets and sweaters hurried along the sidewalk. Some turned into the small shops, while others came out of doorways and joined the stream of pedestrians. The flow of people in motion was ceaseless and dizzying.

Traffic at the corner was stopped, waiting for the light to change. It was the middle of October and leaves still clung to the trees, but a sharp, cold wind blew up from the river. She shivered and hugged her knees tighter. Ten stories up, the miniature city looked as unreal and dreamlike as Tolkien's map of the paths leading to Mount Doom.

Karen Cupnick's ten o'clock art history class was cancelled. This left her with three hours before her next class, and rather than going for coffee with her friends, she decided to return to Williams House and do some reading and eat lunch. Her suitemates should be in class, so she expected to have the apartment to herself until noon.

Karen saw the book bag, purse, and leather coat on the floor as soon as she opened the front door. She recognized them as Nancy's and thought Nancy had probably dropped them there so she could run back to her room and get something. Karen glanced at her watch and saw Nancy had about ten minutes to get to class.

"Nancy!" Karen called out. She expected to see Nancy come tearing down the hall or yell back. But nothing happened.

The suite was so quiet Karen could hear the hum and whine of traffic. The noises seemed unusually loud, and the temperature in the entrance hall was chilly. She decided somebody must have left a window open and turned down the heat. She walked into the living room to check the thermostat.

That's when she saw Nancy.

The window was raised all the way to the top, and Nancy was crouched in the opening. The toes of her running shoes extended over the edge of the sill, and her arms were wrapped around her knees. Her right shoulder leaned against the window's wooden track. The cold wind blew through her fair hair and made it swirl around her face. She looked, Karen thought, like Ophelia staring into the flowing water: beautiful, but remote and unreachable.

"Nancy," Karen said. She tried to keep her voice normal and to tamp down the hysteria she could feel rising in her. Startling Nancy and causing her to fall seemed a genuine possibility. Nancy was precariously balanced on the sill, and it wouldn't take much to upset her equilibrium.

"Yes," Nancy said. She didn't turn her head, and her voice was faint. She seemed to be speaking from a great distance.

"What are you doing?" Karen asked. She considered rushing across the room and grabbing Nancy by the arm. But if she moved too slowly or missed her grip, Nancy might fall. She might even throw herself out.

"Just sitting," Nancy said, in the same bleak tone. "I'm sick of listening to the Agents. I can't stand them any more."

"The agents?" Karen was puzzled. She tried to remember if Nancy had said anything about an agent for the screenplay she'd said she wanted to write. "What agents?"

"The ones running my life," Nancy said. "Ruining my life. Filling me with gold to control me."

Karen didn't understand what Nancy meant, so she kept quiet.

The room was cold, and she didn't know how long Nancy had been squatting on the window sill. If her muscles grew cramped or tired, she might fall without intending to.

Karen made a sudden decision.

She walked over to the window and stood behind Nancy, slightly to her right. Karen said nothing, and when Nancy didn't react, she rested her left hand on Nancy's left shoulder. She was intensely aware of how warm and alive Nancy felt.

"Come inside and tell me about the agents," Karen said. "Art history was cancelled, so I've got time." She paused, then gently squeezed Nancy's shoulder. "I could make coffee."

"I like tea," Nancy said. Her voice was so low she could have been talking to herself.

"Tea's fine," Karen said. She kept her hand on Nancy's shoulder, maintaining a gentle pressure. The wool of the sweater felt rough. "But come and sit while I make it."

Nancy unclasped her knees and grasped the window frame above her head with her right hand. She placed her left hand on the sill, then put her left foot on the floor and followed it with her right. Karen removed her hand from Nancy's shoulder, then pulled down the window. It rattled in its worn tracks. She turned the latch on the window, and relief spread through her like a sudden flush of warmth.

"You know," Karen said, "we ought to skip the tea, and get you over to the University Health Services."

"I'm not sick," Nancy said. She rubbed her face with her hands, then shrugged. She pushed her hair away from her face with her fingers. "Just fatigued by all the gold in my body."

"They can help you with whatever's bothering you," Karen said. "I know where UHS is, so I'll walk over with you."

"Sure," Nancy said. She seemed apathetic and inclined to agree to any suggestion. For that, at least, Karen was glad.

Karen later said that she had found the episode with Nancy deeply disturbing. "I can't say for sure that she was going to jump," she told me when, with Nancy's permission, I talked to her about the events of that day. We met for lunch at a campus cafe about three weeks later, and I asked Karen for her account of what had happened.

"I was afraid that if I so much as touched her, she might fall," Karen said. "But I had to do something. I couldn't just stand there while Nancy sailed out the window." She waved a hand. "That would have haunted me the rest of my life."

Dark-haired and small in size, Karen gave the impression of possessing limitless energy. At twenty, she was only a few months younger than Nancy, but she was someone comfortable with assuming responsibility. She had volunteered at a nursing home during high school and thus had experience dealing with confused and upset people. Even so, Karen found it disturbing that Nancy had seemed so cut off from the world.

"She totally weirded me out," Karen said. She hunched her shoulders and gave an exaggerated shudder. "I was never sure she knew who I was, even when she was talking to me. She was like somebody strung out on drugs." Karen shook her head. "And that stuff about Agents and getting injected with gold...." She leaned across the table, her eyes wide. "I'd never seen *anybody* like that." She frowned. "It's what you imagine it must be like to talk to a snake or some creature from another planet. Just utterly alien and horrifying."

"Do you think you and Nancy will still be friends?" I asked.

On the basis of my impression of Karen, I had expected an immediate yes. She surprised me by taking a minute or so to think about her answer. She finally let out a long sigh and said, "I don't know if that's possible."

She looked at me directly, holding my eyes with hers. "Sure, we can be friendly and go out to dinner and so on." She shook her head. "But I've seen something inside Nancy that I'll never be able to forget. I'll always wonder if the alien part is going to take control of her again." She lowered her eyes. "That's unfair, I know. But it's true."

Karen sat with Nancy at Student Health Services for almost an hour until Dr. Albert Branson, the only psychiatrist on duty, was available to see her. Karen accompanied Nancy into the examination room. She told Dr. Branson about discovering Nancy perched on the windowsill. She also mentioned how, during the last several weeks, Nancy had become withdrawn and moody.

Nancy listened to Karen's version of events with no sign of interest. So far as Nancy was concerned, Karen later said, she could have been describing a scene from a novel. Nancy sat and listened, neither agreeing nor disagreeing with Karen's account. Nor did Nancy volunteer additional information.

"I'm going to leave you here with Dr. Branson," Karen told Nancy. "I'll call your mother and let her know where you are." Karen had met Doris Trail in September when she had come to campus to help Nancy move into the Williams House suite. "Can you give me her number?"

"My mother?" Nancy seemed puzzled. Yet she gave Karen the number, reciting it by rote. Speaking in the same uninflected voice and without being asked, Nancy then provided her mother's office number.

"We will also be talking to your mother," Dr. Branson said after Karen had left. "But first I'd like to ask you a few questions about how your feel now."

Dr. Branson probed only deeply enough to satisfy himself that Karen's account of Nancy's behavior was substantially correct and that Nancy might try to harm herself if she were allowed to return to Williams House.

"I'm transferring you to Lane Hospital, because we're not set up to give you the help you need," Dr. Branson told her. He suddenly smiled. "You may have read about Lane, because it's had many famous patients." Turning serious again, he said, "I was a resident there, so I know you'll get wonderful treatment."

Nancy said nothing. She'd provided terse responses to Dr. Branson's questions about the Agents and the gold injections. She had also admitted to sitting in the window opening, but she'd been equivocal on the matter of whether she had planned to jump out.

Dr. Branson called in one of the nurses to sit with Nancy while he discussed her presenting symptoms with Dr. Garfield Parsons, a psychiatrist at Lane. With Dr. Parsons' concurrence, Dr. Branson made arrangements with a medical transport service to transfer Nancy to Lane. Eastern University's Student Health Center was a drop-in clinic, and people with problems requiring extensive care, whether medical or psychiatric, were routinely transferred to a facility set up to deal with them.

Lane was the best known of the two psychiatric teaching hospitals affiliated with Eastern University Medical School. Although Lane was a leading research institution, it retained signs of its nineteenth-century origins as a private "madhouse." The original building, much modified, was a sprawling federal-style brick structure, with soaring white pillars, a gray slate roof, and cupolas topped with arrow-shaped weathervanes.

The building stood on the crest of a gently sloping hill and was surrounded by several acres of trees. Half a dozen brick cottages were scattered around the main building. Administrative offices and the closed wards were in the main building, and the cottages were halfway houses where improving patients lived with six or eight others under the supervision of a staff member.

Everyone coming to Lane, for whatever reason, was struck by how little it resembled a hospital. It seemed more like the campus of a small, exclusive college. A wrought-iron fence separated the hospital from its suburban neighborhood, but the fence was scarcely three feet high. Thus, the boundary it marked was more psychological than physical. Lane's setting was so peaceful that it was easy to sense the appeal of withdrawing behind its walls to seek refuge from the confusion and demands of the world outside. Despite its transformation into a research hospital, Lane still promised asylum to its patients.

I met Nancy Trail at her initial interview with Dr. Garfield Parsons, the psychiatrist Dr. Branson had called to arrange for Nancy's admission. In addition to working with patients at Lane, Dr. Parsons was also a researcher and a member of Eastern's psychiatry department.

I held a visiting appointment at Eastern Medical School, and Dr. Parsons was helping me acquire the practical experience I needed for my project on psychiatric diagnosis. Mostly this consisted of my sitting in on Dr. Parsons's interviews with patients. After each interview, Dr. Parsons would tell me his tentative diagnosis and explain what led him to it.

Dr. Parsons was a short, spare man in his early forties who seemed in constant motion. He jiggled his legs, patted his hands, rubbed his palms, and ran his fingers through his hair. He would, from time to time and for no apparent reason, frown, grimace, bite at his lower lip, or make a clicking sound with his tongue. The hospital staff joked that the only way you could tell that Dr. Parsons wasn't a patient was that patients weren't allowed to wear ties.

Nancy was brought to the interview room in the main building by a uniformed escort, a woman in a navy blue dress with a white collar. New patients were assigned a room in the closed ward and not permitted access to other parts of the hospital without an escort. The hospital also avoided medicating patients until after they had been assessed and given a tentative diagnosis. Medication might mask or alter behavior that could be an important clue to a diagnosis.

Dr. Parsons leaped out of his chair as soon as Nancy stepped into the room. He smiled broadly, greeted her by her first name, and introduced himself. He then introduced me and mentioned that, if she didn't mind, I'd be sitting with them while they talked. He sounded friendly, although a bit manic, as if meeting her was the best thing he could imagine happening to him that day.

Dr. Parsons made no effort to shake hands with Nancy, but he pulled out a chair at the small, rectangular table. "Why don't you sit here," he said. "I'll sit on the other side so we can see each other. I find it's much easier to talk to people if they're sitting opposite me. Don't you find that too?"

Dr. Parsons's chair was on the side of the table with a button hidden under the edge. The button was wired to a buzzer in the staff room, and at the end of an interview, the psychiatrist could use it to summon an escort.

Depending on how long the button was pushed, it could also be used to signal an emergency. When the crisis call went out, a team trained to deal with violent behavior would rush in and, if necessary, subdue the patient and put him under restraint. This involved wrapping the patient tightly in a blanket to secure his arms and legs. The patient could then be injected with a sedating drug, if the psychiatrist thought it was necessary.

Nancy made no response to Dr. Parson's enthusiastic greeting, but she took the chair he indicated. I sat at the end of the table so I could observe both of them.

The room was bare, with pale green plaster walls and no decorations. The only furniture was the desk-sized mahogany table and four heavy wooden chairs with red leather seats arranged around it. Light came from two white saucer-shaped fixtures suspended from the ceiling by tarnished brass chains. The setup reminded me of a classroom in an old-fashioned elementary school. Yet Lane was one of the most expensive psychiatric hospitals in the nation, with a significant number of famous or socially prominent patients. It was mentioned in dozens of memoirs and biographies.

"Did you eat?" Dr. Parsons asked Nancy abruptly. "Or would you like a drink? I could send someone to the cafeteria for a snack." He pointed at the door and spoke in an urgent manner, as if something crucial depended on her answer.

"I'm fine," Nancy said. Her voice was flat, but she didn't sound upset or worried. She slumped in her chair, her shoulders slightly hunched. I'd noticed when she came into the room that she was taller than average. She was slim, with a taut, well-muscled body. Her hair, uncombed and tangled, was the translucent bronze color of dark amber. It hung to the tops of her shoulders, but she had pushed it behind her ears. This exposed her face, giving her a delicate, vulnerable look. Her eyes were a greenish gray, the irises so dark they appeared black.

She was wearing her own clothes, although she'd been required to surrender her belt. The stiff collar of her white shirt stood out from her neck, and this had the effect of emphasizing her head. This, along with her delicately modeled features and smooth skin, made it easy to imagine her head as a model for a carving on a cameo.

"You're twenty-one, right?" Dr. Parsons said. He was sorting through the papers in a manila folder and didn't seem to expect a response. "You're a senior at Eastern, and your mother lives in New York, which is where you were born. Your father is deceased." He glanced up at her. "Do you have any memory of him?"

Nancy shook her head.

"You were only...what?...four?" Dr. Parsons said. "So that's not a surprise." He looked into the folder. "Your mother told Dr. Branson that your father was prone to serious depression and that he died in a car accident during one of those times. She said that you knew about this. Is that correct?" He looked at Nancy, his head cocked to the side.

Nancy nodded.

"You have no siblings," Dr. Parsons went on. He tapped his fingers on the table while he talked. "You got all your shots as a kid, never been hospitalized. Never been diagnosed with a serious illness and never told you had problems of an emotional or psychological nature." He looked up quickly, as if to catch Nancy by surprise.

Nancy gave no sign that she knew Dr. Parsons was looking at her. She was sitting quietly and letting Dr. Parsons talk. Her eyes didn't seem focused on anything in particular, and she rarely blinked.

"Dr. Branson said you were squatting on the windowsill of your dorm room," Dr. Parsons said. He put down the file. "Your roommate thought you might be about to jump out the window. Is that what you were thinking?"

"Maybe," Nancy said. She sounded disengaged or perhaps extremely weary. "I might have been. I told Karen that I couldn't stand listening to the Agents any more. I wanted them to stop."

"Who are these Agents?" Dr. Parsons asked. He pulled his chair closer the table. He rested his elbows on it, then made a tent of his fingers. "Are they people? People like you and me?"

"Sort of," Nancy said. She seemed bored by the question, and her face remained impassive. "I think of them as people, because I hear them."

"You hear their voices?" Dr. Parsons asked. He leaned toward her, then leaned back. He leaned forward again, as if unable to make up his mind. "Do they talk to you?"

"They tell me things," Nancy said. She was looking toward Dr. Parsons, but her gaze was focused on the table in front of him. "Things I don't like to hear."

"And what might those be?" Dr. Parsons asked. He crossed his arms across his chest, then tucked his hands under his arms.

"Embarrassing things," Nancy said. She shrugged. "But sometimes scary things." She raised her eyes. "I don't want to talk about what they say." Her voice had taken on a sharp edge, hinting that she might be more distressed than her calm surface suggested.

"We don't have to discuss anything you don't want to," Dr. Parsons said. "I mostly wanted to know who these Agents are. But I also need to know if they told you to jump out the window." He uncrossed his arms and put his hands flat on the table. He pushed his chair back, then craned his neck forward to look at Nancy. "Did you hear a voice *ordering* you to jump?"

"Yes," Nancy said. She fluttered her eyelids, then quickly added, "No, not ordering. But Ostra was going on and on, ragging me and making me feel so awful. I wanted him to shut up, because I thought my brain was going to explode."

She paused and her eyes became blank. Then she started talking again. "We studied explosions in chemistry, and we learned they're a form of combustion. Rapid combustion." She nodded, looking directly at Dr. Parsons. "I sometimes think the Agents are going to make me explode. That I might die of too much combustion."

"Ostra didn't order you to jump, but you thought that jumping would keep him quiet." Dr. Parsons spoke as if Nancy had now made clear something that had been puzzling him. He showed no surprise that voices may have been giving Nancy orders or that she worried about exploding.

"I heard all of them on the TV this morning," Nancy said. "Not pictures, only voices. Egon and Tuve, as well as Ostra. But sometimes they whisper softly, and then I hear them in my head." Nancy shrugged, shifting her gaze away from Dr. Parsons. "I was scared when they first started talking to me, but eventually I realized they're my thoughts." She frowned, as if struggling with a math problem. "The Agents are there with me, but they really aren't."

"Have you heard them more often recently?" Dr. Parsons asked. He looked at her intently, as if her answer were of crucial importance. "Is it beginning to get to you?"

"Yeah, it is," Nancy said. "I know something is wrong with me, that I might be going crazy." Her lips trembled. "I can't concentrate on anything, and my life's falling apart."

Her voice was filled with desperation, but only the trembling of her lips suggested any emotion. The whole of her face seemed frozen. It was as if she were drowning, yet not able to distort her perfect features with a cry for help.

"I agree that something is wrong," Dr. Parsons said. He sat still for the first time, and his tone was serious. "But we need to have several talks before we can understand what it is." He leaned across the table and asked gently, "Do you think that might be helpful?"

"I don't know," Nancy said. She had lost her energy, and her voice was dull again. Her mind now seemed somewhere else. "Talking is something I used to do to get people to notice me. I did it in class, and I said good things. Not that anybody could always tell good from bad, of course."

Dr. Parsons let her ramble, then returned to his point.

"Once we get a fix on the problem, we can put you on a drug I'm sure will help," he said. He tapped an index finger on the manila folder. Whether he was emphasizing his point or making a nervous gesture was impossible to tell. "We're going to have to keep you here for a while, though, while we figure things out."

"I can't stay in this hospital," Nancy said. Her voice rose to the edge of panic. "I need to be in my room to deal with the Agents. I need to get away from them."

"We can help you with that," Dr. Parsons said.

"No, you're going to lock me up," Nancy said. She shook her head violently, and her hair fell from behind her ears. "I've been locked up since I got here. Now I want to go to my room."

"Your mother is coming this evening," Dr. Parsons said. "You can talk everything over with her."

"I'm tired," Nancy said. "Really tired." Her voice caught, and she seemed on the verge of tears. "I want to go home and rest, then get on with my life." She brushed back her hair with her right hand, then rubbed her forehead. "I don't want to be here."

"I'm sure you don't," Dr. Parsons said. He spoke in a low, sympathetic voice. "But you're safer with us for a while." He opened the folder and took out a sheet of paper dense with print. He slid it across the table. "Please read this."

"I don't want to." Nancy's head was bowed low, and she twisted the index finger of her left hand in her hair. Then after a moment, she said almost in a whisper, "I didn't tell you about the gold."

"Then please do." Dr. Parsons seemed politely interested.

"The Agents have been injecting me with gold," Nancy said. She whispered the words. "Because it's so heavy, it's made me very tired. I need to go home and take a nap." This was clearly supposed to be the trump card that would permit her to leave.

"A nap is good idea," Dr. Parsons said. He gave no indication that he considered gold injections performed by imagined beings anything out of the ordinary. "But I'd like for you to read this form, and I hope you'll sign it."

Dr. Parsons reached across the table and tapped a finger on the blank line near the bottom of the page. "This is a statement of voluntary commitment. It says you consent to check into this hospital for treatment and agree to stay for a minimum of a week, if your doctors believe it's necessary."

Nancy said nothing. She was no longer toying with her hair, but her eyes seemed unfocused. She gazed at the wall behind Dr. Parsons, and her face was blank and unreadable.

"Take a nap, then get a good sleep tonight," Dr. Parsons said. He spoke as if she were paying close attention to him. "Tomorrow we'll run you through a bunch of tests so we can figure out how to make you feel better." He put his hands together, pressing one against the other, then blew into the crack between them, as if trying to warm his fingers.

"I'm not signing any friggin' paper," Nancy said. "And I'm not staying here." She shook her head again, and the swirl of her amber hair hid her face for a moment.

"Please don't be hasty," Dr. Parsons said. He lifted his hands as if warding off objections. "How about thinking about staying? Maybe give yourself the next hour or so to consider it." Then, lowering his hands, he returned to his offer of food. "I could have coffee and sandwiches sent up to your room. Or ice cream. Whatever you'd like for a snack."

"You have no right to keep me here," Nancy said. She sounded neither agitated nor angry. "I can leave when I want to." She pushed her chair back from the table but remained seated.

"It's possible...," Dr. Parsons said, sounding hesitant. Then he started over. "In my judgment, as well as in Dr. Branson's, you may try to hurt yourself." He spoke the words rapidly, as if it pained him to have to say something that might be taken as an accusation.

"Bullshit!" Nancy said, sounding angry for the first time. "The Agents have stopped talking to me."

"I admit it's only a possibility," Dr. Parsons said, as if agreeing with Nancy. "But as your doctor, I have to take even a remote possibility seriously."

"You can't make me stay," Nancy said. She had turned into a defiant teenager arguing with a parent.

"I wish you would decide to stay on your own," Dr. Parsons said. He spoke firmly. "But if you don't, I can commit you involuntarily to the hospital for up to fifteen days."

"Fifteen days!" Nancy's eyebrows shot up in surprise. She opened her mouth to speak, then closed it again. She jumped up from her chair but remained standing by the table. "Are you telling me I can't walk out of here?"

"Not just now," Dr. Parsons said. He stayed seated, and his voice held a trace of sadness. "But you have the right to ask for a lawyer and challenge the commitment. If you do, you'll have a court hearing, and a judge will then decide if you need to stay here for your own protection."

"What about my mother?" Nancy asked. She sounded both disbelieving and desperate. She grasped the back of the heavy chair, as if she needed to steady herself. "When Mother gets here, she can take me home with her."

"Not at this time." Dr. Parsons sounded sympathetic. "You're an adult, so your mother doesn't have the legal standing to make decisions for you." He leaned across the table and looked up at Nancy. "Not that I won't talk to her and ask her to help us. But I can't release you into her custody."

"Then screw you." Nancy spat out the words. She gave the chair a hard shove, sending it crashing into the table. The chair bounced back and fell over sideways. She kicked out at it as if it were a hand trying to grab her, then started for the door.

I expected Dr. Parsons to push the concealed button to bring out the restraint team, but he kept both hands on top of the table. I wondered if I should offer to help.

"Should I go after her?" I asked. I pushed my chair back from the table.

Dr. Parsons shook his head.

"Nancy," he called out. He was still facing away from her "I know you're angry and upset, but if you stay here with me, we can start getting you back to normal."

"I don't want to talk to you," Nancy said. Her voice was loud and harsh. She stopped at the door, then turned around. "I want to go home."

"Of course you do," Dr. Parsons said. "And before long, you will." He turned to look over his shoulder, then smiled at her. "I can promise you that."

Nancy walked toward the table. She moved slowly and hesitantly, as if she hadn't quite made up her mind to return. I walked around the table and set her chair back on its feet. Without seeming to notice me, she slipped into it from the side.

"I'm not going to sign," Nancy said. She sounded firm, but the anger had evaporated. "The Agents said I would regret it if I put my name on anything." She gave a faint smile. "So you see, I'm not able to sign."

"It's not necessary," Dr. Parsons said. He took back the form he had put in front of her earlier. "I want to keep you here for a few days, so Dr. Branson and I will file the necessary paperwork." He took another form from the manila folder and slid it toward Nancy. "I've already mentioned your rights, but you can also read about them."

"I want to take a nap," Nancy said.

"What a good idea," Dr. Parsons said, sounding enthusiastic. He reached under the table and gave the button a short push. "I wish I could take a nap too, but I'm going to have to stay late this evening."

We sat quietly a moment, then the door opened, and the escort in the blue dress came inside. Her appearance seemed the cue Nancy had been waiting for. Nancy slid out of her chair and met up with the woman in the middle of the room.

Dr. Parsons stood up. "I'm glad to have met you," he said. He spoke as if he and Nancy were at a party, and she had to leave early. "I'll be seeing you tomorrow morning around eleven. Meanwhile, let us know if you need anything."

Nancy stopped and turned around. "Could somebody bring me a Snickers bar and a mug of tea?" she asked. "I know dinner isn't served until six, and I'm quite hungry."

"That we can take care of," Dr. Parsons said.

"I'm sure you will agree...," Dr. Parsons said after the door was closed. He pulled his chair out from the table and dropped into it. He stuck out his legs and put his hands in his pants pockets. "Schizophrenia, paranoid type."

"I don't know enough to agree or disagree," I said. "I know that hallucinations are characteristic of schizophrenia."

"Absolutely," he said. "Not unique to it, but they fit the pattern here. We'll get a more detailed history and rule out organic causes, but with a father with an emotional problem who may have killed himself…many fatal car accidents are disguised suicides.…" He shrugged.

"She realized that the Agents are hallucinations," I said. "Is that common?"

"Not really," Dr. Parsons said. "People sense their minds are the source of what they hear—and it usually is 'hear'; visual hallucinations are rare. But the voices are still real to them, and they usually seem external."

"Is that the symptom that led to your diagnosis?" I asked.

"It's always the most dramatic," Dr. Parsons said. "But did you notice she sometimes started going off on a tangent?"

"Like the talk about explosions?"

"Exactly, exactly," Dr. Parsons said. "And that's what is characteristic of schizophrenia—the fragmentation of thought, the loose associations leading to wild flights."

He halted a moment, then spoke rapidly. "I'm sure you also noticed that Nancy was emotionally unresponsive. She was way beyond low-key. Her emotions were dull or blunted—what we call flattened affect."

"She got angry about being committed," I pointed out.

"She did," Dr. Parsons admitted. "People with paranoid schizophrenia often have appropriate responses. But Nancy's face usually didn't reveal any emotion, her voice was flat, and she didn't make eye contact." He gave me a sly look and smiled. "You seemed a lot more nervous than she did. Particularly when she mentioned the Agents."

I felt myself blush. "Are hallucinations always threatening or unpleasant?" I asked.

"Yeah, mostly," Dr. Parsons said. "It's stuff people don't admit they have problems with. Like sex and anger. They try desperately to keep their feelings under control, but they break out of the corral like wild stallions."

Dr. Parsons gave a sudden whoop of laugher. "Listen to me sounding like John Wayne, when the closest I've been to a horse was at a pony ride when I was eight." He suddenly sat bolt upright for no apparent reason. "But you get the idea."

"The Agents are hallucinations?" I asked. "Technically, I mean."

"The voices are hallucinations," Dr. Parsons said. "False perceptions. But the gold injections, that's a delusion—a false belief." He raised a hand. "Don't ask why some false beliefs are delusions and others are just wrong. But as a rule of thumb, delusions involve misinterpreting facts, like Nancy thinking she's being injected with gold because her mother was."

He straightened his tie. "The hallucinations, the delusions, the disordered thinking, the flattened emotions, the lack of motivation and loss of interest in her classes...they all add up to somebody cut off from reality." He sucked on his lower lip a moment. "That's what we call a psychosis. Schizophrenia is one of them, and manic-depressive illness is the other."

"Do you think she wanted to kill herself?" I asked.

Dr. Parsons walked around to the other side of the table. "I don't know if 'want' is appropriate," he said. "She's so cut off from the world, it doesn't make sense to talk about her intentions." He raised his right hand and pointed his index finger toward the ceiling. "What's not in doubt is that she was responding to a 'command' hallucination."

I looked puzzled. Dr. Parsons said, "Ostra gave her an order. He told her to jump." He looked at me and frowned. "Like the Son of Sam's dog telling him to shoot people?"

"Oh, right," I said. Dr. Parsons was standing behind the chair where Nancy had sat. "She's very young."

"Prime schizophrenia time," Dr. Parsons said. "Late teens, early twenties, maybe the thirties, and hardly ever after forty. Hence the old name, *dementia praecox*—deranged thinking in youth."

He sat in what I thought of as Nancy's chair. "I'll most likely start her on Thorazine," he said. "A high dose at the beginning to knock out the hallucinations, then bring it down."

"What chance of recovery does she have?" I asked.

"If we can find a drug that works and she sticks to it, we can restore her pretty much to normal functioning for a year," Dr. Parsons said. "After that, it's the rule of thirds. A third get better and stay better, and a third relapse from time to time." He stopped, then, sounding reluctant, he said, "The last third stay crazy most of the time." He gave a small, humorless smile. "And we don't know which third Nancy will be in."

I visited Nancy during her third week at Lane. She had moved to a cottage and was allowed to walk around the grounds, take part in organized activities, and visit the snack bar. She wasn't permitted to leave the grounds, and she was scheduled to meet twice daily for an hour-long session with her psychiatrist.

I met her at Livingston Cottage one afternoon, and we walked along an asphalt path that skirted the main building and led to the tennis courts at the back. It was November, and the weather was considerably colder. The trees scattered around the grounds had lost their leaves, and their branches formed complicated patterns of erratic lines against the sky.

"How do you like it here?" I asked.

Nancy was dressed, as before, in jeans and running shoes, except now she was also wearing a black down jacket. Her amber-colored hair was pulled away from her face and tied in a ponytail. The cold wind gave her cheeks a rosy glow, and she seemed the picture of health.

"I feel safe here," she said. She spoke hesitantly, as if having to think hard to find the words. "Outside, I was scared all the time, even though I don't know if I realized what it was."

She didn't look at me, and even while she was talking, her features lacked animation. Her face seemed fuller, and she had lost the lithe, athletic look I had noticed when she entered the interview room. Her motions seemed slower.

"Did you decide to stay more than a week?" I asked.

"You mean did I go to court and lose?" Nancy's voice, for the first time, held a trace of amusement. "No. Once I was on medication and going to therapy, I got better. I then voluntarily agreed to stay for up to a month."

Nancy walked slowly and with an awkward gait, as if hobbling on a sprained ankle. Dr. Parsons had mentioned that muscle stiffness, as well as weight gain and sedation, were side effects of Thorazine. The drug's drawbacks were obvious, yet it had reestablished Nancy's connection with reality. She might feel slow and heavy, but she no longer believed it was because the Agents had injected her with gold.

"I'm hoping to return to classes next week," Nancy said. She sounded uncertain. "Karen brought me my books, and I'm able to study now. I can't catch up in developmental psych, but I'm okay in the other courses."

"Are you looking forward to getting back?" I asked.

"I think so," Nancy said. She stopped and turned toward me. "I'm a little scared," she said. "I don't know how people will react, and I don't know how I'll react." She looked at me and gave a nervous smile. "I've never been crazy before."

I tried to keep the shock from registering on my face. Was I suppose to deny that she was crazy? I didn't think so. I decided she was being brave. She was calling herself crazy so she could get used to the idea that, even when she got out of Lane, her problems wouldn't be over.

She was facing a lifetime of struggle, even if she were in the luckier two-thirds. She likely would reach a time when she could no longer tolerate the sedating effects of the drug. She might skip a few doses, then stop completely. She would then feel fully alive, awake, and responsive for the first time in months. Going back to the state of having your feelings blunted by chemicals working on your mind would seem abhorrent.

But the voices would eventually return. The Agents would once more mock and belittle her. They might again command her to kill herself, and next time someone like Karen might not be around. If Nancy survived, she would have to return to Lane or someplace like it. She would be given an antipsychotic drug, perhaps a new one, required to attend psychotherapy sessions, and made to follow a sort of summer-camp routine of meals, therapy, and rest.

The cycle would then start over.

"I think we should walk back," Nancy suddenly said. She wrapped her arms around her chest. "It's colder than I expected."

"We can go to the snack bar," I said. "Don't you like tea?"

"I do," she said. "And it doesn't have to be decaffeinated. Dr. Parsons says tea doesn't have enough caffeine to interfere with the action of my medicine." She gave me a tired smile. "Any good news is welcome, you know."

I returned her smile and glanced up. The sky directly above us was clear, but it had the raw, scraped color of a fall day. The sun shining in at a low angle was weak and failed to provide the golden light needed to wash away the grayness.

"Winter is coming," Nancy said, following my gaze upward.

Forget about Nancy Trail for a moment and consider the following imaginary case.

Suppose your friend Dave lives alone in a house only a couple of blocks from yours, and late one afternoon you decide to drop by to see him. When you arrive at his house, Dave is sitting on the front porch with a small chainsaw beside his chair.

"Hey, good to see you," Dave says. "I've decided to cut off my left leg with this chainsaw, and you can be my audience."

"If that's a joke, I don't find it funny," you tell him.

"I'm not joking," Dave says. "I've been thinking about it off and on for a couple of months, and I've decided it's the right thing to do."

"But why?" you ask, wondering if Dave can be sincere. "What can you possibly hope to accomplish by mutilating yourself?"

"I want to call attention to the problems of amputees and other movement-impaired people in our society," Dave says. "I'm a committed advocate for the disabled, and I want to show my solidarity with them by joining their ranks."

"So you plan to mutilate yourself as a publicity stunt?" You are finding it hard to believe Dave is serious and not stringing you along for his amusement.

"You could call it that," Dave says. "But I consider it a form of political action. I want the public to see how my life changes when I stop being able to walk in the easy, unaided way that I do now."

"Advocates for children don't think they have to join the ranks of children," you say. "And animal-rights advocates don't have to start acting like dogs."

"Well, nobody can become a child or a nonhuman animal," Dave says. "So the issue doesn't arise for them. But I can become disabled."

"You can, but why should you?" you ask. "You could find many other ways of calling attention to the problems of the impaired that are much less drastic."

"But that's the point, don't you see?" Dave says. "Amputating my perfectly normal leg is so drastic, so over the top, that it's the most effective political statement I could possibly make."

"And you think it's worth it?"

"Yeah, I do," Dave says. "I wouldn't ask anybody else to do it, but it's right for me."

"There's a social cost," you know. "It's not only you who's paying a price. Society has to use its resources to patch you up, make you a prosthetic leg, and rehabilitate you. Those resources could go to someone who lost a leg in an accident or a war."

"I take the point," Dave says. "But I'm counting on my action making more resources available to every disabled person. So my costs will ultimately result in a net gain." He smiles. "Besides, free expression always comes at a price. You wouldn't stop a radical from speaking because of the cost of police protection, would you?"

"You realize that your leg isn't going to grow back?"

"Of course," Dave says, looking amused. "I assure you that I'm not delusional. I also know I'm taking a chance of dying from blood loss and infection. I'll tie a tourniquet around my thigh and call 911, and I won't make the cut until I hear the ambulance coming down my street."

You remain quiet a moment, unsure what to make of Dave and his intention. He appears rational in discussing his proposed amputation. He seems to have thought the matter through and to understand the consequences of the action. Even so, what Dave intends to do seems completely crazy.

You take out your cell phone and call 911.

TWO QUESTIONS

If, in response to your call, the police come and take Dave to the emergency room of a hospital, the psychiatrist called in to examine Dave will need to answer two questions: (1) Is the patient competent to make important, life-altering decisions? (2) Is the patient likely to be a threat to himself or others?

Because Dave is explicit and vocal about his plans, the answer to the second question is obvious. Yet even Dave's announced intention to harm himself isn't sufficient, legally or morally, to justify a decision to commit him to a psychiatric unit against his will. If his plan were to harm somebody else, commitment would be justifiable, but when a patient's intention is to act in way that harms only himself, how the first question is answered determines whether he should be detained.

Our society recognizes and protects the autonomy of the individual. We allow people to shape their own lives by the decisions they make. Some who have spent ten or twelve years training to be a neurosurgeon decide that the life is not for them and drop out to become painters. Others quit their jobs as hedge-fund managers to teach kindergarten.

People also make decisions that are less far-reaching but that involve serious risks. Those who crave excitement may take up base jumping and leap from the top of the Petronas Tower, attempt to climb K2 without supplemental oxygen, or hunt for new species of poisonous snakes in the highland jungles of Malaysia. These activities are much riskier than such ordinary activities as playing tennis, shopping at a supermarket, or solving crossword puzzles on the subway.

Many of us may believe that people who "throw away" their careers as neurosurgeons or "stupidly" risk their lives to achieve momentary thrills are "crazy." We may try to persuade our friends not to give up their careers or risk their lives, but we don't lock them up to stop them from carrying out their plans. We may even have grudging admiration for their willingness to resist the pleadings and condemnations of their friends, defy social expectations, and pursue their dreams. They are, after all, only exercising their autonomy and deciding how they want to live their lives.

Which is why Dave presents such a problem for the psychiatrist (call him Dr. Good) who examines him. Dave answers Dr. Good's questions in the same frank way he answered yours.

So far as Dr. Good can judge, Dave isn't delusional and doesn't show signs of a psychotic disorder. He gives intelligent answers to the questions Dr. Good asks, and his emotional responses (including anger at having been brought in for psychiatric assessment) seem appropriate.

Yet Dave continues to insist that he plans to cut off his leg. Does this mean Dave is crazy? It's enough to make anyone suspect he might be, but to assume that it proves he is begs the question. Dr. Good would need independent evidence.

Also, Dr. Good knows that people have performed self-destructive acts of political protest throughout history. Some have immolated themselves to object to the Vietnam War, and some have become suicide bombers to oppose the American presence in Iraq. Thousands of others have sacrificed themselves for their religious or political beliefs. Thomas More, for example, could have escaped the headman's axe by rejecting the authority of the Pope and signing an oath of loyalty to Henry VIII. He chose to die.

Dr. Good's psychiatric examination of Dave doesn't turn up anything to suggest Dave isn't competent to make significant decisions. But the examination was brief, and if Dr. Good had more information, he might either conclude that his initial impression was wrong or confirm his view that Dave is competent.

No one could fault Dr. Good for refusing to hospitalize Dave. If Dave were planning to do something like getting married or buying a house, his act would have serious consequences, but if the action turned out to be a mistake, most of them would be reversible. Dave is planning to perform an act, however, that is extreme and life-threatening and has irreversible consequences. Thus, Dr. Good would be justified in erring on the side of caution and committing Dave for observation and testing.

If Dr. Good and his colleagues find that Dave is competent to make autonomous decisions, he will be released within a week. If they decide to release him, their answer to Question 1 is that Dave is competent, and their answer to Question 2 is that he poses a danger to himself. (For someone found to be incompetent and to pose a danger to others, consider the case of Tatiana Tarasoff in the second part of the chapter titled "The Boy Who Was Addicted to Pain.") If Dave still wants to carry out his self-amputation, he is now free to do so.

Dr. Good doesn't prevent Dave from acting for a paternalistic reason ("it's not in his best interest") or in the service of a compelling state interest ("He's trying to avoid Army deployment orders"). Dave's autonomy has been restricted. But the restriction is for a short time, and it's only to make sure he's capable of acting autonomously. Delaying a political statement for a week or two isn't, prima facie, as serious as allowing a mentally incompetent man to cut off his leg.

HELP NEEDED

When most people hear the phrase "involuntary civil commitment," they tend to think of cases almost as extreme as Dave's. They are likely to conjure up some half-remembered scene from *One Flew Over the Cuckoo's Nest* and imagine a raving lunatic trussed in a straitjacket being dragged down to a padded cell by burly attendants.

The truth is less dramatic. Most often, the person committed involuntarily to a psychiatric ward is someone who, like Nancy Trail, has done or said something to indicate that she needs help coping with the immediate pressures of life. The indication may be as flagrant as someone slashing her wrists with a razor blade or washing down a dozen tablets of OxyContin with a pint of scotch. Yet it may also be as indirect and comparatively undramatic as someone saying he is tired of his life and doesn't care if he wakes up the next morning.

QUESTION 1

State laws vary on procedures for committing patients against their will and on involving a civil court to rule on the legitimacy of the confinement. In all cases, however, a doctor, usually a psychiatrist, is required to review the available evidence to answer the two questions mentioned above: Is the patient competent? Is the patient a threat to himself or others?

To answer these questions, the psychiatrist isn't required to make a specific diagnosis. Sometimes a diagnosis is possible, sometimes it isn't. A psychiatrist would like to have the luxury of dealing with a patient who has had a thorough medical examination and psychiatric work-up. These would include a neurological exam and laboratory and imaging studies to look for organic disorders such as a drug reaction, epilepsy, a brain tumor, or an endocrine disorder, as well as psychological tests like the Beck Depression Inventory or the Minnesota Multiphasic Personality Inventory. (Dave will get such a work-up, if Dr. Good commits him.)

Most often, a psychiatrist must make a decision about a patient's competence in a crisis situation using only fragmentary evidence. In the case of Nancy Trail, Dr. Branson must be guided by Karen's observations and his impression of Nancy, and Dr. Parsons must make a decision on a similar basis. Both psychiatrists are called on to exercise their clinical judgment about the likely behavior of someone they have little information about. There is no test comparable to a brain scan or a blood count that will allow them to determine if Nancy is competent.

Their clinical experience and relevant knowledge makes it likely that their judgment will be more reliable than that of others, even other physicians. Dr. Parsons, for example, recognizes that Nancy displays symptoms of a psychotic disorder and severe depression, both of which are risk factors for suicide. The command hallucination (the voice ordering her to jump) makes the chance of suicide even greater. Also, suicide is responsible for 30 percent of the deaths of college students, and 20 percent of emergency medical care is suicide-related. Thus, Dr. Parsons would have good statistical reasons to think Nancy might be a danger to herself, even if she didn't display the symptoms of two psychiatric disorders.

As the case of Dave shows, the likelihood that Nancy may act in a way that might result in her harming or killing herself isn't a sufficient reason to commit her to a psychiatric ward. The crucial element in Dr. Parsons' judgment is that Nancy lacks the mental competence to make decisions about her own welfare.

Because of her psychiatric disorders, Nancy's autonomy must be considered impaired. Her delusional beliefs about the Agents and their activities make it impossible for her come to grips with the world. She cannot weigh and assess evidence about her options, then make choices about how she wants to live her life. What she does or decides to do is, in a real sense, not of her own willing.

Nancy is not exercising her autonomy, because her autonomy has been captured, caged, and rendered powerless by the disorders that have taken control of her. It's as if the alien invaders in *Invasion of the Body Snatchers* have replaced Nancy with a replica that looks like her but has been altered in ways that don't allow her to express her will. The commands she responds to are now those of her disordered mind. The answer to Question 1 for Nancy Trail is: no, she's not competent to make serious decisions.

QUESTION 2

Is Nancy a threat to herself—is she suicidal?

She herself couldn't be sure, but crouching on the sill of an open window on a cold day was behavior that struck Karen as strange. More important, it wasn't something she'd seen Nancy do before—it wasn't characteristic of Nancy's usual behavior. When Nancy then began to rail against the "Agents" that were "filling" her body with gold, Karen felt certain that something was wrong with Nancy. She decided Nancy was in danger of killing herself, whether she intended to or not.

Dr. Branson and Dr. Parsons arrived at the same conclusion. Unlike Karen, they had the legal power to detain Nancy against her will. This is an awesome power, because it allows a physician to deprive a patient of her liberty. He can do so for only a comparatively short time (days, not months) and with judicial checks in place. Yet this isn't the sort of power we allow even the police to exercise. We reject "preventive detention," the practice of locking up people we suspect might do something illegal, and the police can interfere with our liberty only if we are suspected of having committed a crime.

Why do we grant physicians the power to lock us up and study us for a while? We do so because we realize that we need some sort of fail-safe mechanism to keep us from injuring ourselves or others when our minds are out of balance and we aren't able to reason properly. We depend on physicians to assess our competence in using information to make decisions.

When the system works properly, as it did with Nancy Trail, it can save lives. When it fails, as it did in the case of Cho Seung-Hui, who in 2007 shot to death thirty-two people and wounded twenty-five others at Virginia Tech, the result can be disastrous.

FOUR

Unsuitable

R OB HARDY WAS only twenty, but he looked like death.
He was lying in a hospital bed with the mattress tilted at a
sharp angle so he was almost sitting upright. His face, smooth
as a young child's, was slack, and his hazel eyes, half-shut, were dull
and blank. His short tar-black hair stuck up in feathery tufts, and a
silver tracery of dried drool followed a curving path from a corner of
his mouth to the tip of his pointed chin.

Rob's skin, even on his neck and hands, had the dusky, lusterless
burnt-orange color of preserved tangerine peel. This was the most
obvious sign that he was in liver failure.

A plastic line dripped a colorless solution into the needle taped into
place in his right arm, and leads from a heart monitor and an oximeter
curled across the sheet, connecting him to the machines behind his
bed. A pale blue tube clipped to his nostrils supplied extra oxygen. A
third plastic line ran from under the sheet to a urine-collecting bag
attached to the side of the bed. The bag's sparse contents were the color
of dark copper.

Rob seemed barely aware of the group gathered around his bed.
Dr. Sara Gibbs, a hepatologist, was making morning rounds with six
or eight residents. I was along because I was an outside observer of the
committee recommending which liver-failure patients should become
candidates for transplants.

The shortage of donor livers was so desperate that allocating one to
somebody unwilling to take care of it was tantamount to jeopardizing
the life of the next person in line, who might die while waiting. Given
the high stakes, I made it a point to visit potential candidates likely to

become subjects of dispute within the committee. I wanted to form my own opinion of them and not have to rely solely on the reports of social workers. Otherwise, I wouldn't be able to judge the fairness of the process.

"Rob, how are you feeling?" Dr. Gibbs asked, putting her hand on his arm. She bent over to look at the way the IV needle was taped down, then checked the line to see if the drip was working properly. She flicked her finger against it, and a drop slid down the slim tube.

"Horrible." Rob mumbled the word, not looking at her or anyone else. His eyes remained half-closed and unfocused.

"Want some ice?" Dr. Gibbs asked.

Without waiting for the answer, Dr. Gibbs scooped a few chips from the insulated bucket on the bedside table and held the spoon to Rob's lips. He craned his neck forward and sucked the ice off the tip of the spoon. Exhausted by the effort, he fell back onto his pillow. His eyelids flickered, then closed.

"What can you tell us about this patient, Dr. Felix?" Dr. Gibbs asked.

A plump, dark man looking hardly older than the young man in the bed slipped an index card from under the rubber band of a bundle. He stood at the head of the bed, facing Dr. Gibbs and his fellow residents.

"Mr. Hardy is a twenty-year-old Caucasian male admitted to the hospital at about two A.M. this morning with fulminate liver failure due to acute hepatocellular necrosis." Dr. Felix's voice got creaky, then returned to normal. "Apparently caused by the ingestion of perhaps twenty-five grams of acetaminophen after the consumption of a large but undetermined amount of alcohol during a binge of social drinking."

He squinted at the card. "The acetaminophen was seemingly self-administered as five-hundred-milligram tablets. While white residue, presumed to be dissolved tablets, was recovered on gastric lavage, significant hepatic damage had already occurred."

The resident continued his report in the formulaic way favored by academic medicine. I knew from my medical-history reading that this style of medical reporting had originated in the Paris hospitals of the eighteenth century and was a sign that medicine had become scientific. The patient, for the purpose of diagnosis, was viewed not as a unique individual, but as a case fitting into a disease category. The case, then as now, was characterized by a statement of the patient's symptoms and medical history, along with the findings of the physician's examination.

Laboratory tests might be added if the doctor thought they might produce helpful data.

Dr. Felix ran through Rob Hardy's numbers—temperature, blood pressure, respiration, BUN, hematocrit, bilirubin, SGOT, and prothrombin time—then reported on his heart rhythm. I paid little attention to the recital or to Dr. Gibbs's questions about changes and trends. Whatever the numbers showed, I knew Rob was extremely sick and might die at any time. He had, at best, only a day or so left without a new liver.

The liver is the only vital organ without a temporary substitute. Someone whose kidneys shut down can be sustained by dialysis, ventilators can supply oxygen, and even somebody whose heart fails can be kept alive by a small, partly implanted, ventricular pump that keeps the blood flowing. But nothing can take the place of a liver except another liver.

I had heard Rob Hardy's story from George Muzorsky, the gastrointestinal Fellow on call when Rob was brought in by the EMS. Rob was in his junior year at a nearby state college and had passed out during a party at the apartment of some friends. The party was large and noisy, and no one had noticed Rob's absence. Then someone looking for the bathroom found Rob on the bedroom floor. His friends shook him and splashed water on his face. He eventually woke up, but he seemed disoriented. His friends decided he was very drunk, but otherwise okay.

Rob had his own apartment, but he was so out of it that two of his friends decided to take him to his mother's house. Rob was more awake when they arrived at around two A.M. "Too much vodka," he told his mother. "My head is really, really killing me. I gotta get to bed."

Caroline Hardy had seen her son drunk before and wasn't worried, but his stepfather, Alan Manners, was. Caroline and Alan had been married for less than a year, and Alan hardly knew Rob. Alan suggested taking Rob to a hospital emergency room, but Caroline said that would be an overreaction. Rob, insisting that he didn't need help, managed to stagger up the stairs to sleep in his old room. Caroline and Alan went back to bed.

Rob was not awake when his mother and stepfather left for work at seven-thirty the next morning. *He'll sleep it off,* Caroline thought, *but he's not going to escape a lecture.*

Rob was still in bed when Caroline and Alan returned at around six that evening. They shook him and called his name but were unable to rouse him. Alan went into Rob's bathroom to wet a washcloth and noticed an empty Tylenol bottle on the floor. Fragments of foil nearby suggested it had just been opened. Rob's mother, shocked and frightened, called 911, and Rob, still unconscious, was rushed to the ER in an ambulance.

He was intubated, his stomach was washed out, and he was given an IV infusion of acetylcyusteine, a substance that can prevent acetaminophen from causing liver damage. To be most effective, the drug must be given within ten to twelve hours after the acetaminophen is ingested, and no one was sure when Rob took the Tylenol. Was it before he went to bed? During the night? Sometime the next day? Rob couldn't say.

"I'm guessing it was when he got upstairs," George told me. "He had a monster headache, and we didn't recover much residue. If he swallowed the tablets at two A.M., and we saw him in the ER at seven P.M., that's seventeen hours. Acetylcyusteine usually does zilch after sixteen hours. But there's no downside to trying it, so we're giving it to him IV." George frowned. "He's lost so much liver, I don't expect him to survive."

"Did he say anything?" I asked.

"Not a lot. We asked him to tell us how much he swallowed." George shook his head. "He wasn't sure. We got a psychiatric consult, but Dr. Denning said he would only admit to having a headache and taking something for it."

I went back to see Rob in the early afternoon. I accompanied Bruce Wong, a social worker whose job was to interview Rob and report to the transplant committee.

Bruce wore a short white coat and, typically, a silly tie. Today his tie was printed to look like a rainbow trout, its head and glassy eye at the bottom. Bruce was overweight and low-key. Patients liked him and would talk to him. Ironically, he wasn't given to empathy and didn't seem to notice when his probings caused pain. He also had no sense of humor, despite the ties.

The IV in Rob's arm was still dripping the colorless fluid, and the same lines were connected to the monitors behind the bed. The bag collecting urine was now a quarter full, and I noticed how closely the oxidized-copper tint of the fluid matched the color of Rob's face and arms.

Bruce told Rob who he was, then introduced me. Rob, with obvious effort, held out his hand to shake. I gave his hand a gentle squeeze and felt the barest pressure in return. His skin was dry and sandpapery, and his flesh felt boneless and heavy. Bruce sat by the foot of the bed.

I took the remaining chair, which was about where Dr. Felix had been standing when he presented Rob's case.

"Did you think the pills would harm you?" Bruce bluntly asked. "The Tylenol?"

"I wasn't thinking," Rob said. His voice was weak and he still looked very sick, but he was more alert. He paused to inhale oxygen, sniffing at the tube. "I was, like, in a dream and my head was killing me and I wanted it to stop. I found the bottle and swallowed some capsules. Maybe a handful? Maybe a dozen? I didn't count."

"Were you planning on hurting yourself?" Bruce asked.

"I wasn't planning anything," Rob said, his voice stronger. His eyes were now fully open, their dark irises surrounded by halos of pale yellow.

I felt a surge of sympathy. Maybe, for a moment, Rob had wanted to kill himself. But now he was afraid he was going to die, and he didn't want to. He couldn't undo the impulse, but maybe he could escape its consequences.

"I was totally out of it until I woke up in the hospital," Rob said. He licked his lips. "Somebody told me I'd swallowed a bunch of pills and asked me what they were." He shivered, squeezing his eyes shut. "I didn't understand at first."

"Have you recently had trouble sleeping?" Bruce's voice was impassive. The questions sounded like they came straight from a textbook. "Have you worried about doing something to harm yourself or other people?"

"No," Rob said, with some force. "No to everything. I'm a math major, and I've always got too much homework, and that stresses me out. But I have fun with my friends." His tone turned plaintive as he added, "I'm happy with my life."

Bruce asked Rob if he had "relationship problems." How upset had he been when a girlfriend broke up with him and moved to Colorado? Did he have "issues" with his mother or stepfather? Did he drink a lot or use cocaine, methamphetamines, or other recreational drugs?

Rob answered in a level voice, but he got more agitated as the questioning went on. His breathing became rapid, and he stirred in the bed, shifting his feet under the sheet. He scratched at one arm, then the other. The motion caused the IV line to swing in a short arc.

"Aren't you about done?" Rob asked. "I'm feeling sort of nauseated." His voice was thick and slow. He closed his eyes and snorted at the oxygen. "Dr. Gibbs says I need a new liver and that's all that can help me." He licked at his dry lips.

I got up and scooped some chipped ice out of the bucket the way I'd seen Dr. Gibbs do. Without saying anything, I held the spoon for Rob, and he opened his mouth like a baby bird. He swallowed the chips whole, then licked his lips again.

"I'm scared," Rob said. His voice was blurred and so faint that only I could hear him. But he didn't seem to be talking to me. Or to anybody, except maybe himself.

Then his eyes suddenly turned wild and staring, their orange-white irises making them look inflamed. Tears trembled at the corners of his lower lids, then trickled down his cheeks. I pulled a tissue from the box on the bed tray and put it into his hand. I squeezed his arm gently, but I could think of nothing to say that wouldn't sound either fatuous or false.

"Bear with me," Bruce said stolidly, glancing down at his clipboard. "I've got to finish this form."

I laid a hand on Rob's leg for a moment, then excused myself from the room. I couldn't do anything to help him, and Bruce had his job to do. Besides, I had formed my opinion.

The ten people making up the transplant committee met later that afternoon and discussed Rob Hardy's case for forty minutes. Most cases took half that time. The central question was whether Rob was likely to commit suicide and thus waste a donor liver.

The psychologist thought Rob was in denial about depression, but Dr. Denning's consult note and Bruce's psychosocial evaluation supported the notion that the Tylenol episode was due to drunken confusion. That Rob was a high-performing college student and had the support of a stable family counted in his favor.

A few committee members thought Rob had tried to kill himself. But he was only twenty with a long life ahead of him, and everyone agreed this earned him the benefit of any doubt. Rob was approved for a liver transplant, with the condition that he agree to counseling about alcohol abuse.

The transplant coordinator, following the directions of Rob's physicians, entered him on the United Network for Organ Sharing (UNOS) waiting list as Status One. This meant he was being kept alive only by intensive medical care and was expected to die within twenty-four hours unless he received a liver transplant.

Rob's status put him at the top of the list, but his getting a transplant depended on the death of somebody with a compatible blood type, a healthy liver, and a family willing to donate it. The donor also had to die within a limited geographical area; organs deteriorate when cut off from their blood supply, and refrigeration and drugs can preserve them for only a few hours. So many contingencies were involved that Rob could die before getting a liver. He probably would.

I stayed after the meeting to listen as Susan Worth, the transplant coordinator, explained how UNOS operates to Caroline Hardy and Alan Manners. The three sat at a small round table in a corner of the transplant office. I sat nearby, but I had explained when I introduced myself that I was there only as an observer.

"So all we can do is wait?" Caroline Hardy asked after Susan had finished her explanation. Caroline spoke each word with exaggerated enunciation, as if she had learned English as a foreign language.

Caroline was short and slight, with curled hair colored an unnatural shade of black. She had such a school-teacherish manner that I hadn't been surprised to learn from Rob's social-situation note that she was the head of the English Department at a suburban middle school. Alan was the principal of the high school across the street.

"I'm going to issue a pager to each of you," Susan said. This was part of the upbeat talk she gave to the family of every transplant candidate. "We have no way of knowing when a liver for Rob might become available. But I'll page you when we've got one, and then you can come back for the surgery."

"Surely we can speed up the process," Alan said. His gold wire-framed glasses with round lenses seemed embedded in his fleshy face, and this made me think of Professor Umrat in *The Blue Angel*. "If Rob was Mickey Mantle, I bet you could find a way of getting him a liver today."

"Not really." Susan's voice remained bright, but now it had a sharp edge.

"Don't upset yourself, Alan," Caroline said. "Please." She frowned and touched his cheek. "It won't help any of us."

"Should I call Senator Teague?" Alan fixed Susan with a hard stare. "My nephew is his chief aide, and I consider the Senator a friend." He was like a man ordering dinner at a restaurant and not expecting to be denied what he wanted.

"Call the Senator if you want to," Susan said, making a joke of the question. "But not even the President can pull an organ donor out of the air." She turned to Caroline. "I did want to mention one other possibility."

"Something besides a transplant?" Caroline held her hand over her heart, and I could hear the hope in her voice.

"Nothing like that," Susan said. "But a living-donor transplant. One of you might be able to donate a part of your liver to Rob, if you decided to."

Susan paused, then glanced from Caroline to Alan. "We could schedule the surgery, and you wouldn't have to worry about getting a liver in time to help Rob." She gave a wide smile, as if offering the two a surprise gift.

"I had no idea that was possible." Caroline's perfect enunciation had disappeared. She stood up. "He can have a piece of my liver. Of course!"

"Don't rush ahead," Alan said harshly. He crossed his arms over his chest and cocked his head at Caroline. "You've got high blood pressure, and the surgery might kill you."

"But I'm his mother," Caroline said. She gave him a stern look, daring him to disagree. "And now he needs me."

"But I need you too." Alan's voice was pleading. Caroline sat down, and Alan rested a hand on her shoulder. It could have been a gesture of affection or a claim of ownership.

"I've just got to do it," Caroline said. She pounded her fist into her open hand and looked up at Susan.

Alan compressed his lips. He looked like he was biting back words, but he said nothing. He took his hand off Caroline's shoulder and slouched in his chair.

"Wait a minute," Susan said. She held up her clipboard. "You're both running ahead without any facts. A donor has to be in good health and have normal liver function. And although your liver grows back in a few weeks, you can die from the surgery, as well as suffer complications."

"How many people die?" Alan brightened with interest.

"About two percent," Susan said. "But thirty percent develop complications, and a few get so sick they need a transplant themselves." She fixed her gaze on Caroline. "If you're interested, you can talk with Dr. Rossiter."

"Of course we're interested," Caroline said.

"We've got to have all the facts," Alan said. He tugged at the knot on his tie, tightening it to cover his collar button. "It's not my practice to go into anything blindly."

"You'll each have a chance to talk with Dr. Rossiter privately, then decide if you want to go to the next stage," Susan said. She held up an index finger. "And one more thing. You must have a blood group compatible with Rob's, which is O. Do you know your blood types?"

"I'm an A, and Alan is an O," Caroline said without hesitation. "We donate blood at school every year." She smiled and her eyes sought Susan's. "Is mine okay?"

"Sorry," Susan said. "O recipients can only have O donors. So you're not eligible, but Mr. Manners is."

"Damn!" Caroline spat out the word. She turned to her husband. "I don't think you should donate. You and Rob aren't particularly close."

"But you and I are," Alan said. He touched her arm. "I want to do whatever is necessary, whatever needs to be done."

"It's wonderful of you," Caroline said. Her voice was soft, barely audible. "But I don't want to put you on the spot." Then she clenched her fists and said in a tight, frustrated tone, "If only it could be me. I feel so worthless."

"I'll talk to this doctor," Alan said. He slapped the table. "I'm in good shape, and I'll be a fine donor. Rob's going to be okay." He forced a smile, then frowned.

Caroline gave her husband a look of gratitude, her eyes shining with tears. I decided I had misjudged Alan. He was a stuffed shirt and pompous, and I thought he was going to balk at becoming a donor. Yet here he was displaying a kind of casual heroism that was impossible not to admire.

"After Dr. Rossiter assesses you, you'll need to have some tests and sign a consent document," Susan said. "But let me page him and see if he's free."

Dr. Larry Rossiter came down to the transplant center, nodded to Susan and me, then introduced himself to Alan and Caroline. He was a bluff, hearty man with curly black hair and an olive complexion. The starch had gone out of his white coat, and it was wrinkled across the rear.

"Come into the exam room and let me look you over," Dr. Rossiter said to Alan. He took a manila folder from Susan, then put his hand on Alan's shoulder and guided him down the hall toward a room with an open door.

"Keep your fingers crossed." Alan turned and glanced at Caroline. Looking calm, maybe even happy, he gave her salute and a little smile.

Ten minutes later, Alan came out of the room looking dejected. His shoulders were slumped, his head bowed, and he walked toward Caroline with slow, dragging steps.

"I'm sorry, really, really sorry," he said to Caroline. His voice was soft, and he didn't meet her eyes. "Dr. Rossiter says I'm unsuitable as a donor."

"Oh, Alan," Caroline said, sounding anguished. She hugged him, then pulled back and looked into his face. "Did he say why? Don't tell me there's something wrong with you too. I couldn't stand it." She seemed on the verge of tears, and her pretentious pronunciation had evaporated completely.

"No, I'm fine," Alan said. "It's something about the shape or location of my liver." He looked mournful. "But I feel so terrible that I can't help Rob."

"Oh, I know you do." Caroline spoke in almost a wail. "I feel the same way. But I'm so glad that you're okay. Now we have to pray that Rob gets a liver in time."

I watched the two of them move toward the elevator. They walked together, Alan with his arm around his wife's shoulders, hugging her to him.

I felt sorry for Rob Hardy and sorry for his mother. Where Alan was concerned, my feelings were mixed. He wasn't being candid with Caroline. "Unsuitable" was the word the transplant center used to protect people from family reprisals when they admitted to a physician in private that they didn't want to become donors. So Alan's disappointment was counterfeit. He wasn't prepared to give up a part of his liver for his stepson, but he didn't want Caroline to know it.

If personal autonomy means anything, it means we get to control what is done to our bodies. The point of informed consent in medicine is that it protects this autonomy. A doctor needs to get our permission even to check our pulse. So if someone doesn't want to submit to surgery and risk disaster, including death, by giving up part of his liver, he should be free to refuse. What's more, he shouldn't have to fear any negative consequences for his refusal. If he does, this violates his autonomy, because his consent can't be free.

This is exactly what I witnessed with Alan.

Alan, without the protections of consent and confidentiality, might have felt forced to undergo surgery because of pressure from Caroline. Or, if he had openly refused to donate part of his liver to Rob, his relationship with Caroline would have been damaged, perhaps destroyed. She now saw him turned down for reasons he apparently couldn't control.

The center's consent process had worked the way it was supposed to. Yet I was bothered by Alan's behavior. With the help of the process, he was allowing Caroline to believe he was unacceptable as a donor for physical reasons. Maybe such a deception was necessary to protect his autonomy, and maybe protecting autonomy was important enough to justify it. I could accept that.

What disturbed me was Alan's presenting himself as a willing, even eager, liver donor. His enthusiasm hadn't kicked in until Susan had said he and Caroline would *each* be having a private talk with Dr. Rossiter. Only then did his manner become downright heroic. I hadn't realized it at the time, but his enthusiasm had been as bogus as his dejection at being declared "medically unsuitable."

Alan was quicker, bolder, and much slyer than I had suspected. He'd found a way to pervert a process designed to protect him into one to achieve his own aims. His wife not only didn't blame him for not being a donor; she was grateful to him for his outspoken willingness.

I felt naive for ever having thought Alan was brave. Naive and annoyed, because he had duped me as well as his wife.

Rob Hardy received a donor liver early the next morning. The surgery took place without a hitch, and by late afternoon he was sitting up in bed looking almost normal. His skin's yellow tinge was fading, and he was able to smile. His new liver was holding death at bay.

Alan Manners visited Rob every day he was in the hospital.

Ethical principles require that people do the right thing but don't require them to become moral heroes. It would have been decent, kind, and generous of Alan Manners to donate part of his liver to his stepson. We would all admire him for such a display of courage and benevolence.

Yet Alan had no obligation to contribute a liver segment to anybody, including Rob. Even though Alan had good reason to believe Rob would die unless he got a transplant quickly, Alan's refusal to become a donor wasn't wrong. Rob needed part of Alan's liver, but he had no legitimate claim to it. That Alan is his stepfather makes no difference. Alan's decision may have been ungenerous and uncourageous, but it wasn't wrong.

We have a general duty to act benevolently and help others, but we get to decide when and how much help we want to offer. If we decide the cost is too high in a particular case, we don't have to help. It would be extremely generous of me to give my Cartier watch to the next destitute person I meet. I have no obligation to do that, though, and giving away the watch wouldn't cause physical pain or even take courage—except maybe the courage to tell my wife what I'd done.

Alan was required to give his informed consent before becoming a donor. The purpose of informed consent is to protect autonomy, and to achieve this, consent involves more than saying yes or no. Alan had to be told exactly what he would be getting into if he agreed to become a donor. He had to hear about the pain of surgery, the chances of infection, the possibility of short-term and long-term medical complications, and the risk of death. Alan also had to learn how much good his donation was likely to do for Rob, as well as what would happen if Rob failed to get a transplant quickly. In short, Alan needed to be given enough information to make an informed decision.

Equally important, Alan had to be given sufficient time to deliberate. Rather than being rushed into making a decision, he needed time think about Rob's situation, Caroline's worry about her son, and the consequences Rob's death might have for his marriage. He had to consider, especially, what his relationship with Caroline would be like if he decided not to donate a liver segment and Rob died. Alan also needed to consider the risks and suffering involved in becoming a donor and reflect on his own desires, fears, and values. He could then either agree or decline to become a donor.

The transplant coordinator and Dr. Rossiter, we can assume, provided Alan with sufficient medical data to make a decision. The criteria for informed consent were thus met, and Alan decided not to become a donor.

He wasn't candid about his decision in talking to Caroline afterward. Indeed, he encouraged her to believe that he couldn't become a donor for reasons beyond his control. No one at the transplant center told her otherwise. It's thus reasonable to ask if the transplant center wasn't engaging in deception in failing to reveal to Caroline that, even though Alan was medically qualified to become a donor, he had decided against it.

The blunt answer is no.

A center establishes criteria that must be satisfied before someone becomes a liver-segment donor. Most criteria are medical, and a candidate can be excluded on such grounds as having an incompatible blood type (as with Rob's mother), a hepatitis infection, uncontrolled hypertension, or some serious or debilitating disease. When it comes to ruling out potential donors, medical criteria usually cause few problems.

(Not everyone accepts being ruled out without strenuous protest, and some insist on becoming donors against medical advice. These are what transplant surgeons call "heroic donors." Parents, in particular, may demand that they be allowed to save the life of their child, even when doctors have warned them they are putting their own lives at risk. Should parents be allowed to exercise their autonomy and do what they can to help their child? Perhaps so, but this is a different issue.)

Those closely associated with someone needing a transplant don't always want to become donors. This fundamental fact causes a major problem when it comes to making sure that the informed consent of a donor is genuine.

Thomas Starzl, a surgeon who pioneered both liver and kidney transplants, eventually decided not to use living donors, because his experience convinced him that the weakest or least valued member of a family is always targeted as the donor. Others in the family deliberately or unconsciously manipulate that person into "volunteering."

To prevent people from being railroaded into becoming donors, a transplant center needs a procedure that gives volunteers the chance to refuse to become a donor without having to suffer reprisals from family members or friends.

The transplant center where Alan was evaluated made sure that volunteers had the chance to discuss their decision out of the earshot of others. They were assured, in private, that if they wanted to change their minds (for any reason or no reason at all), they were free to do so. Because a "suitable" donor must be one whose consent is both informed and freely given, if a volunteer decided he didn't want to become a donor, he would be considered an "unsuitable" candidate. Others wouldn't be told exactly what made him unsuitable. Protecting confidential information by refusing to reveal it is not deceptive in any way.

Here, then, was the mechanism that protected Alan's autonomy. It protected him from the fear that if he didn't become a donor against his wishes, his refusal to help Rob might result in his alienating Caroline and destroying his marriage.

The transplant center's mechanism did what it was supposed to do. That it didn't turn Alan into a forthright and admirable person doesn't mean the system had a flaw. Some may think that the flaw was in Alan.

FIVE

Nothing Personal

"**D**R. MUNSON? MAY I speak to you a minute?"

She was tall and willowy and blonde. Perhaps in her late thirties, about my age at the time. She was quietly beautiful, with high cheekbones, a narrow face, and a softly rounded chin. Her eyes were the slate-blue color of the sky on a rainy summer day.

She wore a tailored gray suit that fit so perfectly she might have been modeling it. A black polished leather handbag hung from a strap over her shoulder, and her earrings were small cut stones with the deep red glow of rubies. She was the epitome of the impeccably, expensively turned out young executive. Even so, I found it easy to imagine her in jeans and a T-shirt with her honey-colored hair tied back in a ribbon. Her natural style seemed casual and athletic, more Abercrombie than Brooks Brothers.

"Sure," I said. "Have a seat." I gestured toward the chair across from me at the small square table.

I was eating lunch in the cafeteria of one of the National Institutes of Health (NIH) hospitals. It was an anonymous place, with greenish fluorescent light, blond wood furniture, and a pervasive steam-table odor. Some diners were patients in bathrobes and slippers, sitting alone or huddled over tables with their visiting families. Others were physicians, scientists, nurses, and various sorts of administrators. I was the outsider, brought in to consult with a regulatory committee charged with deciding if ovarian cancer patients should have access to experimental drugs without being in a clinical trial.

"I'm Karlina Miller," she said. "Call me Kar."

Her smile was dazzling. She put down the cup of hot water she was carrying and held out her hand. Half rising, I took it in mine. Her skin was warm and soft, but her grip was strong, almost aggressive.

I tried to recall if I had seen her at the morning session. Thirty or so NIH staff members had crowded around the seminar table, and Tom Thorgood, the committee Chair, was the only one I knew. Yet I was certain Kar hadn't been in the room. She was so striking that I would have noticed her. Everyone would have noticed her.

Kar slipped into the chair, putting her purse on the one beside her. Without saying another word, she concentrated on tearing the foil wrapper off her tea bag, then dipped the bag into her cup. She was like a child absorbed in play and oblivious of her surroundings. The water, not hot enough to draw the tea properly, turned a pale gold.

Keeping an eye on Kar, I went back to eating my soup, but her beauty and her polish made me self-conscious. Every movement I made seemed clumsy, and I felt as awkward as an adolescent. I was careful not to slurp the soup, and I repeatedly dabbed at my mouth with the paper napkin.

Kar took a sip of the weak tea, then set the cup down and wrinkled her nose. "Agh, that is genuinely vile," she said. She laughed in what seemed to be spontaneous good humor, and I laughed with her. We could have been lovers being silly together, ready to find everything funny. She said, "I don't guess the coffee is any better."

"Dunkin' Donuts has nothing to worry about." I nodded toward my cup, still smiling. Our remarks were pointless and almost random. We were taking the measure of each other, but we were also putting off the moment when Kar would have to say why she wanted to talk to me.

"I'm executive editor of Criterion Books." She spoke hurriedly, as if unaccustomed to having to identify herself and wanting to get done with it. "I'm responsible for nonfiction, mostly general interest books, but some science and medicine."

"So you publish those books on alien abductions and how to choose your cosmetic surgeon?" I gave a quick smile to show her I was teasing and resumed eating my soup.

I thought I now knew why she wanted to talk to me. I expected her to ask if I had an idea for a popular book on medicine and ethics. I began to run through various projects I might propose.

"Hmmm, aliens are a bit passé." Kar echoed my facetious tone, and her eyes brightened with fun. "But, hey, cosmetic surgery is always hot." Then her face became serious. "I took the train down from New York last night so I could be here for today's committee meeting."

"Right," I said uncertainly and waited for her to go on.

She sipped her tea a moment, looking at me over the rim of the cup. The rubies in her ears flashed as she tilted her head. I saw that her manicured nails were bitten ragged at the edges. It was the first flaw I'd seen in her polished veneer.

"I can't believe that a clinical trial appeals to a popular publisher," I said. I was puzzled and wondering where she was headed. "I don't think you'd sell many books."

"I have ovarian cancer." Kar's voice was soft and quiet, and she kept her eyes focused on mine. I wanted to look away, but I didn't. Perhaps deliberately, she had ended our charade of casual conversation.

"I'm sorry to hear that." My voice was strained, and my stomach tightened. I put down my spoon. A small, pale green pool of soup remained at the bottom of the bowl, but I wasn't going to be able to finish it.

"Epithelial, stage four. High grade," Kar said, still speaking softly, almost as if to herself. She was telling me that she was in serious trouble. Stage IV means the cancer has spread into the abdomen or even beyond, and tumors classified as high-grade are those that are most aggressive. Then, giving the words an ironic inflection, she said, "My doctor turned out not to be any better at diagnosis than I was."

"What do you mean?"

"I didn't have dramatic symptoms," Kar said. "Only minor abdominal pains and feeling swollen. Irregular periods." She gave a humorless smile. "But my periods were always irregular, and I've had stomach pains since first grade."

"Did you see your doctor right away?" I asked.

A part of me didn't want to know anything about Kar's illness, yet another wanted to hear the details. Also, I sensed that, for whatever reason, she was making me a gift of her story, and I couldn't turn away from her. She was a stranger, but we had laughed together for a moment, and I felt compelled to listen.

"Oh, I saw her all right," Kar said. "I got a pelvic exam, and dear old Dr. Soutz couldn't find the tiniest little thing wrong with me. Probably irritable bowel syndrome, all perfectly harmless. Blah-blah-blah." I heard fierce anger in her sarcasm. "She gave me a low-dose antidepressant for the cramping. I took it for maybe three months, thinking I was just fine."

Kar stopped talking, but her expression remained somber. She picked up the sodden tea bag and devoted her attention to dipping it repeatedly into her cup of yellowish water. Tea slopped over the rim and into the saucer, but the color in the cup didn't noticeably darken.

"The pains became sharper, and I started getting nauseated and throwing up," Kar said. The distress showing on her face humanized her beauty and made her more appealing. She was somebody you wanted to help. "I was scared, so I went to a gynecologist, who took my symptoms seriously. My CA-125 level was outrageously high, and a CT scan showed I had swollen lymph nodes behind my abdomen." Her voice became strained, and the slate blue of her eyes seemed to grow darker, clouded by the memory.

Her gynecologist had done the right tests. CA-125 is a substance that increases in the blood when ovarian cancer cells multiply, but it can also increase for other reasons, so a high level is not definitive evidence of cancer. Yet when an increase is accompanied by scans showing pelvic abnormalities, ovarian cancer becomes the most likely diagnosis. To settle any doubts, surgeons may insert a fiber optic instrument through a small incision near the navel and look inside the abdomen.

"Did you have a laparoscopy?" I would have considered the question intrusive if Kar hadn't so plainly wanted to tell me her experiences.

"I was way beyond that." Kar hunched forward, looked down, and twisted her cup in its saucer, making a grating noise. "I had wide-open abdominal surgery. I lost my ovaries, my uterus, and part of my intestine. Cancer was in a lot of places, and my surgeon said it was very 'sticky.'" She raised her eyes to meet mine and held my gaze. "He couldn't get it all."

Kar compressed her lips and shut her eyes. Her eyelids looked as thin and delicate as tissue paper, and below their surface tiny blue veins formed a random pattern. After a moment, she opened her eyes and sat up straight in her chair. She was visibly pulling herself together.

"That must have been a hard time," I said.

My voice was strained and hoarse. What she had said earlier hadn't prepared me adequately, and I was shocked. Her cancer would most likely kill her within a few months or a year, yet I hadn't read that fact in her face. Because of her beauty and charm, I had put her down as someone for whom things had always gone right. And maybe that had been so. Now, though, things had gone horribly wrong.

"It's still a hard time." She wasn't whining, only stating a matter of fact.

"I can understand why." I wanted to reassure her, tell her everything was going to be all right. But I couldn't do that, because it would be an obvious lie. I asked, "Did you have chemotherapy after the surgery?"

"Radiation first, then cisplatin, carboplatinum, and Taxol," Kar said. "I wouldn't wish cisplatin on Hitler." She shook her head and rolled her eyes toward the ceiling. "It makes you nauseated and they pump you full of fluids, so you have to throw up and pee at the same time."

"I'm sure it's terrible," I said. The words were clumsy, and my sympathy came out as formal and cold.

"And it didn't work." Kar's voice turned harsh, and I sensed her fear. "That's what I hate most about it. You figure if you're going to be filleted like a goddamned fish and then poisoned with chemicals, you ought to get something out of the deal. Something like staying alive."

Kar stopped talking, then made a quick wiping motion with one hand, as if cleaning off a blackboard. "That's unfair," she said. The harshness had dropped away, and her tone was again conversational. "I got ten months of mostly normal life, and that was enough time to get used to not having a fiancé anymore."

"You broke off your engagement?"

"Not exactly." Kar's lips curled into a tight, twisted smile that marred her beauty for a moment. "And Sam didn't break it off either. He sort of just disappeared during my chemo; then he sent me an e-mail saying cancer was too upsetting for him to deal with and wishing me the best of luck. I tried calling him, but he never answered." Kar gave an abrupt laugh. "I took the hint and haven't heard from him since."

"Even long marriages can break up when a spouse gets seriously sick," I said. "The other can't take it."

"I sort of understand that," Kar said. "But cancer's not contagious, for God's sake. You can talk to people face to face."

"He behaved badly," I said.

"And he was the love of my life," Kar said, without irony. We sat silently, listening to the white noise of chatter and the occasional sharp clink of plates and silverware. Kar's face was relaxed again, and her eyes had a distant look. Coming out of her reverie, she shrugged, then smiled ruefully. "I wish I could ignore cancer, but I'm scheduled for more chemo. Different drugs this time."

"At least you know what to expect." I took a swallow of coffee. It was tepid and bitter, with an acid, medicinal flavor. "Are you being treated here?"

"No." She flicked the hair off her forehead with a quick gesture. "But that may change." She leaned across the table as if about to tell me a secret, and I caught a whiff of her light, floral perfume. She said, almost in a whisper, "I've always had a dread of talking to strangers."

"But you've been talking to me." I leaned forward so that our foreheads were almost touching. The smell of the perfume was stronger, and I sensed the warmth coming from her. "It's easy enough, isn't it?"

I was surprised. With her beauty and poise, as well as her tailored gray suit, she came across as supremely confident, even intimidating.

"No, this is very hard for me," Kar said, her voice still low and whispery. "I can force myself to do it when I have to, but it's a struggle." She sat up, and so did I. Our moment of physical closeness was over.

Kar said, "I asked the secretary who the ethics consultant was, and when you came out of the meeting room, she pointed you out." She smiled, and her cheeks dimpled at the corners of her mouth. "What a relief. I think I was expecting some shriveled-up priest who'd give me a dirty look and tell me to get lost."

"I save my dirty looks for the people who sit in front of me on airplanes and recline their seats," I said. Kar's smile was wide enough to show her dimples, which made me feel I'd said something witty. I was glad the tone was lightening, but I was puzzled. Exactly why had she told me her story?

Kar's smile faded, and a long minute passed. When she said nothing, I asked, "Why did you want to meet me?" I picked up my cup, then put it down again. The coffee would be cold and disgusting. Besides, my stomach still felt as tight as a clenched fist.

"I need to talk to you about an ethical problem." Kar frowned in concentration, putting her thoughts in order. "I read about a new drug called Ovcax that's said to be better than anything now used in treating metastatic ovarian cancer." I nodded to show I was following her. "But Ovcax isn't FDA approved, and my doctor says that the only possible way to get it is by enrolling in the NCI clinical trial."

"Right," I said, nodding. "That's the trial our committee is discussing. Dr. Joel Chandler is the P.I. [principal investigator]. Have you talked to him?"

"I saw him last week," Kar said. "I should buy a season's ticket on the Metroliner." She smiled, making her cheeks dimple. The strain lines disappeared from her face, and she looked radiant. Then the smile evaporated as quickly as it had come. "Dr. Chandler was very nice, and I qualified for his study. But the trial compares treatment using the standard drugs with those drugs plus Ovcax. That means I might be randomly assigned to the group that doesn't get Ovcax."

"That's right," I said, nodding. I glanced at my watch and saw the meeting was starting in twenty minutes. I wanted coffee to keep me alert during the long afternoon. I didn't have enough time to wait in line for it and also to listen to Kar. But not listening to Kar was no longer an option.

"To be in the trial, I have to go to the hospital once a week and have my blood drawn so they can test it for an increase in liver enzymes," Kar said. "You don't have to do that if you're getting the standard drugs from your own doctor."

"They need to know if Ovcax causes liver damage," I said, thinking she didn't grasp the experimental setup. "Even if you're not in the group getting Ovcax, they need to check you, because the study is double-blinded."

"I understand that." Kar sounded impatient. "They also do a tumor biopsy after you get your last treatment, and I understand that. They want to see if the drugs have altered the tumors."

"Did you sign up for the study?"

"No," Kar said. "That's the ethical problem." She uncrossed her arms. "I can only get Ovcax by agreeing to the blood draws and the biopsies." She drummed her fingers on the table while she spoke.

"Okay," I shrugged. "But what's the ethical problem?"

"You don't see it?" Kar frowned deeply, furrowing her forehead "The problem is that the protocol is coercive and violates my rights as a patient. Don't you agree?" She looked at me earnestly, then hurried on. "I'm forced to submit to Dr. Chandler's rules, and there's no guarantee even then that I'll be treated with Ovcax. I only have a fifty-fifty chance."

"Wait a minute." I held up a hand. "I understand what you're saying, but that's the issue we'll be deciding at this afternoon's meeting. I'm not supposed to express an opinion until the decision is made."

"I don't expect you to commit yourself." Kar spoke quickly, and her eyes were open wide, almost staring. Once more I saw how frightened she was. "Can't I tell you what I think, the way anybody can? At least anybody bold enough to approach a total stranger, then talk his ear off." She gave a small laugh, and some of her desperation seemed to ease. "You know who I felt like?"

I shook my head.

"The Ancient Mariner," she said. "Telling my terrible story to somebody who doesn't want to hear it, but is too polite to make me go away."

"That's not me," I said. "I don't mind listening."

I realized as I spoke that I was telling the truth. Kar and I weren't friends, but we might have been. I was beguiled by her beauty, without question; but I was also taken with the straightforward, often ironic way she talked about her problems, avoiding self-pity and hiding her fear. We now shared a kind of intimacy, a closeness of feeling that was almost like affection.

"That's nice of you." Kar bobbed her head in a slight bow, then returned to her topic. "Do you understand my point? Ovcax is my last chance, and I must take it. NIH is being unfair in withholding it and making me become a guinea pig just to have a chance of getting it."

"I see what you're saying." To give myself a chance to think, I took a sip of the cold, bitter coffee. "But maybe you shouldn't attach so much importance to Ovcax. It has no track record of success, and the great majority of experimental drugs don't do any better than the standard therapy."

"I know, I know," Kar said, sounding exasperated. "But it might work, and I want to try it." She looked at me, a film of tears making her eyes a deeper blue. "Isn't this a free country? Can't I do whatever I want, so long as I don't hurt anybody else?"

"That's something the committee will have to decide." I turned my wrist to look at my watch. "Speaking of which, I've got about four minutes to get back."

"I'll walk with you, if you don't mind," Kar said. "But first let me give you my card." She opened her purse and took out a thin silver case engraved with the striations of Tiffany's machine design. "Call me anytime you like," she said, handing me the card. "I wrote my home number on the back."

"Thanks." I tucked the card into the inside pocket of my blazer. I was going to offer my own card, then let the moment pass. "I won't be free to tell you about the meeting."

"I understand," Kar said. "But we could just talk." She forced a laugh. "I'm in an excellent position to give you the patient's perspective on ovarian cancer treatments."

Leaving the cafeteria, we strolled across the charmless concrete plaza that joined the buildings in the complex. The sun was out, but it was March, and the weather was gray and raw.

"I'm not sorry to be leaving." Kar hunched her shoulders and crossed her arms over her chest. "I don't know why research hospitals have to be so dreary. If this were like a resort, patients would have a better chance of recovering."

"I believe it has something to do with money," I said. "Research takes priority over patients' comforts."

"But is it cheaper to make everything so ugly?"

I held the door for Kar, and we entered the building where the meetings were held. The space seemed devoted wholly to offices, but chemical odors drifted through the corridor, so labs had to be located somewhere. Kar's heels clicked on the vinyl floor, and its polished surface reflected vague and shifting images of the two of us. We looked like ghosts in a horror story, trapped in a spirit world and unable to return to the place where we had spent our lives.

"I appreciate your talking to me." Kar put her hand on my upper arm and gave it a gentle squeeze. The gesture was casual, and I cautioned myself not to make it into anything more. "I'm afraid I ruined your lunch."

"The cafeteria did that before I met you," I said. She laughed. "But I doubt talking to me did you much good."

"Oh, but it did." Kar's voice rose with insistence, and her eyes widened, revealing a paler shade of blue in the changed light. "I wasn't kidding about the patient's perspective. I feel much better knowing that you'll be thinking about people like me during the discussion."

We arrived at the elevator, and I pressed the button. The red indicator light showed the car was on the fifth floor and coming down. We stood together awkwardly, shoulder to shoulder. We were neither complete strangers nor friends.

"Are you going to wait around for the committee's decision?" I kept my eyes on the indicator light. "My guess is that we'll be here until six o'clock, maybe later."

"I'm returning to New York," Kar said. "I'll call Dr. Chandler tomorrow and find out if they're going to let me have the drug. I'm keeping my fingers crossed." She sounded like a child praying for the impossible, for something to make the bad stuff go away and allow life to return to normal.

"I don't know how things will go," I said, shaking my head. I had been careful to sound neutral before, but Kar's ingenuous hope now made me feel false. I needed to hint at what might happen. "You really mustn't count on it."

"Ron, I have to count on something." Kar said it simply, holding my eyes with hers.

She hadn't used my first name before, and I experienced a thrill at hearing it. But I also felt uneasy. By naming me, Kar had turned me into a person, someone able to care about what happened to her. I had lost my status as a virtually anonymous professional, someone imported to play a role for the day, and I found that disturbing.

I wanted to say something encouraging, but the elevator arrived before I could think of anything. We stepped aside to allow the crowd of people inside to get off.

"I'll say goodbye here," Kar said. I turned toward her, and she held out her hand. "Thanks for your kindness, and maybe we'll meet again."

"Under better circumstances, I hope." I squeezed her hand, and our fingers touched for longer than they had before. "Maybe after your troubles are behind you."

She nodded and smiled broadly, showing the dimples at the corners of her mouth. She gave me a small wave as I stepped into the elevator. By the time the doors slid shut, Kar had turned and was walking back down the hall.

Because I was the consultant, I gave the only formal talk. The record, I said, supports the policy that drugs should be available to everyone only when proved safe and effective by clinical trials. We departed from this policy when we treated millions of postmenopausal women with hormones, and the outcome was that the women didn't get the benefits we expected and their risks of breast cancer and heart disease increased significantly.

The question we are asked to resolve, I said, is whether clinical trials are inherently coercive. Every trial protocol has rules patients must accept to be admitted and gain access to the test drug, and these rules might be seen as infringing the right of patients to be self-determining.

A clinical trial would be coercive, I argued, if the only way patients could get the treatment recognized as the current best was by volunteering for it. But women with ovarian cancer can get this "standard of care" by going to their own physicians. With respect to Ovcax, we have no reason to think it's superior to the standard of care, so patients lack a reasonable basis for claiming that gaining access to it is crucial. They may believe it's crucial, and if they do, they are free to volunteer for a clinical trial, but no one is forcing them to.

So far as a protocol's rules are concerned, patients are told in advance how the trial will work, what will be expected of them, and what risks they will run. It's then up to them to decide whether the chance of getting treated with Ovcax is worth the disadvantages. The point of requiring informed consent from participants in a clinical trial is to allow patients to be self-determining in just this way.

During my talk and in the question period afterward, I spoke against the view Kar had expressed. Some committee members agreed with her as completely as I disagreed, and I used all my skills in reasoning to convince them they were wrong. My position prevailed in the end, and the committee voted to continue denying patients access to test drugs unless they were enrolled in clinical trials and belonged to the experimental group.

I had won. At the moment the committee's decision was announced, I felt smug with triumph and congratulated myself on achieving the result I had wanted. The committee was adjourned without a date set for the next meeting.

I shook hands with people and chatted for a few minutes about the discussion with Tom Thorgood, who had brought me in as a consultant. Tom had to leave to make his evening rounds at the hospital, so the conversation was brief. We said goodbye, and I left to catch the airport shuttle.

I waited at the shuttle stop outside the hospital's main entrance. The weather was still raw and chill. The sun was sinking low, and twilight cloaked the quadrangle's slab-sided buildings in subdued shades of rose and purple, softening their harsh lines. The traffic noise from the road in front of the complex was like the steady roar of a fast stream rushing though a narrow, rocky gorge.

I was exhausted by the long day, and my mind was running out of gear. Putting my hand into my jacket pocket to check for my airplane ticket, my fingers brushed the card Kar had given me. I felt a surge of the sort of sadness and pity you feel when you see news footage of the victims of a bomb blast. You wish the world were a place where such terrible things didn't happen, but don't see anything useful you can do about it.

The shuttle came into view, its engine roaring and its headlights slashing through the gloom of the fading light. I dropped Kar's card into the steel-mesh trash basket beside the shuttle sign. I felt an immediate sense of relief, as if I had disposed of evidence linking me to a crime. My lingering feeling of guilt was inappropriate, but that didn't make it less real.

Karlina Miller's wanting to try an experimental drug she thinks might help her is reasonable. She has a life-threatening disease that standard therapies haven't brought under control. She appears doomed to an early death unless some new but untested drug halts the disease. Even if the drug is ineffective, what has she lost? She's no worse off, and she has the satisfaction of knowing that she gave defeating the disease her best shot.

This is a powerful argument in favor of removing the restraints on the autonomy of individuals with respect to access to treatments.

Ultimately, though, I don't think it's powerful enough to justify setting aside the protections provided by the drug regulation system.

FDA POLICY

The way things work, even someone in Karlina Miller's situation can't gain access to an unapproved drug. The Federal Drug Administration, the primary agency regulating drugs, has decided that if a drug hasn't been tested for safety and effectiveness, it can't be prescribed for anyone. Indeed, even before a drug is tested in a clinical trial, it must have FDA approval for investigational use, and the application for approval must include data about the comparative safety of the drug's use in test animals. (The only exception to the FDA's rule involves patients who are excluded from a clinical trial of a drug due to such factors as age or having additional medical problems. If there is no standard therapy recognized as effective, the company making the drug can apply for "expanded access" use, and the drug can be administered to this select group.)

Experience supports the value of the FDA's tough testing policy. Doctors and patients, as it happens, are not good judges of whether a new treatment is better than a standard treatment. We now know, for example, that the value of hormone replacement therapy (HRT) for postmenopausal women is offset by a significant increase in the risk of heart disease and breast cancer. Before a clinical trial of HRT was conducted, the medical community was enthusiastic about its benefits (decrease in mood swings and hot flashes, increase in sex drive, etc.) and encouraged many women to give it a try. Some doctors even thought that HRT *reduced* heart disease. Such cases could be multiplied. (Read "It Seemed Like a Good Idea" on page 155 for a detailed example.) But the reasoning behind the FDA policy is obvious enough: clinical trials are the best way we have to identify treatments most likely to be safe and effective.

FREEDOM FROM THE FDA

Under ordinary circumstances, no one wants to be exposed to unknown risks as a result of taking a drug. The situation may be different, however, for people who have been told that they have a fatal illness for which approved therapies are likely to be of limited effectiveness. Perhaps people unfortunate enough to be in such extraordinary circumstances ought to be given access to whatever drug they might like to try.

That some people should be permitted to experiment on themselves seems prima facie reasonable. The FDA's role is to protect the public from ineffective and unsafe drugs. Yet it seems absurd for the FDA to try to protect people who are beyond protection and want only the freedom to attempt to save their lives.

Here's a possibility: We could pass a law allowing people diagnosed with an illness expected to be terminal to have any drug they want administered to them by a physician. The drug wouldn't need to be approved by the FDA or even recognized as an experimental drug supported by animal studies. Thus, anyone wishing to use a drug merely rumored to be effective would be free to do so.

To avoid such obvious pitfalls as the problems caused by a mistaken diagnosis, we could fine-tune the law. We could require, for example, that the diagnosis be confirmed independently by two physicians and that the disease be expected to be fatal within, say, six months to a year. (Medicine is good at diagnosing fatal diseases, but much less good at predicting exactly when, in individual cases, they will cause death.)

YES, BUT...

Such legislation would protect the autonomy of fatally ill individuals, but it would also have profound negative consequences. The most serious of these is that the law would threaten the integrity and effectiveness of our medical-care system, which is science-based and thus supported to a considerable extent by FDA drug regulations. If any and all drugs were allowed for even restricted therapeutic use, the rise of various forms of quackery would be encouraged.

It's easy to imagine some physicians deciding to open clinics devoted to providing terminally ill patients with the treatments they want. Thus, clinics offering "experimental drugs for liver failure" or specializing in the treatment of "hopeless cancer patients" could be expected to spring up all over the country.

The existence of such clinics would encourage patients to place their trust in worthless but well-publicized drugs. Those who might derive benefits from established therapies (and many do) would be likely to turn away from them in favor of unproved remedies. The very fact of legalization would give an air of legitimacy to virtually any kind of treatment for the fatally ill. Those desperately attempting to lengthen their lives might actually shorten them.

Neither private insurance nor government programs would be likely to pay for treatments with untested drugs. We can imagine that desperation might drive fatally ill people into exhausting their savings, mortgaging their houses, and leaving their families destitute as they search in vain for a miracle drug that will halt their disease.

People made desperate by disease already seek treatment by quacks, spend their savings, and die leaving their families penniless. The clinics offering these experimental cures, however, are usually in Mexico, Thailand, Brazil, Costa Rica, or other places where no FDA-like agency upholds science-based medicine. That other countries allow desperate people to bet on a long shot at considerable cost is not an argument for following their example.

Placing restrictions on the therapy an individual is free to choose can be construed as limiting autonomy. We ordinarily assume that people want to protect themselves from therapies judged by accepted scientific standards as useless, harmful, or unproven. Thus, we aren't inclined to complain that drug regulations are an unwarranted violation of an individual's freedom to choose. (Who would choose to be treated with an ineffective or dangerous drug?)

From the viewpoint of someone like Karlina Miller, however, such regulations are blatantly paternalistic. We want to protect her and others like her from exploitation, even if they are willing to be exploited. We want to protect their spouses and their children (when they exist) from suffering the secondary consequences of quackery.

Most important, though, we want to protect the rest of us, and we are willing to violate the autonomy of the terminally ill to achieve this end. We might, with our imagined law, succeed in guaranteeing freedom of choice in the case of terminally ill patients, but this could result in subjecting the overwhelming majority of the population to the unacceptable risks that might result from compromising the drug-approval process.

We can't guarantee that dying people can have the drugs they want. The salve for this wound to personal autonomy is that the evidence shows that the chance of an experimental and untested drug being effective is slight.

SIX

"He's Had Enough"

I HAD NEVER SEEN such a horrible wound.

The right half of Walter Post's face was missing. His cheek was cut away, and his upper and lower jaw had been removed at the midline. Stainless-steel wire bound the remnants of his jaws together. His tongue was a raw stub of flesh protruding from a dark clot that seemed to fill what was left of his mouth.

I stood at the foot of the bed in the cramped surgical intensive care room with Walt's wife Martha and his ENT surgeon, Dr. Roy Stone. Michael, the SICU night nurse, a thin blonde man in his thirties, used metal tongs to pull out the pads of blood-soaked gauze packed into Walt's wound. With the gauze gone, the bone of the floor of his skull was visible as an irregular white patch in a field of red. Blood was seeping from the edges of multiple surgical wounds.

Martha gripped the metal frame at the end of the narrow bed, then turned her face toward the blank yellow wall. Her plump jaw was clenched, and her skin was the pale gray-white color of ashes, yet she didn't cry. Then, with what seemed to require an effort of will, she looked back at Walt.

Walt's devastated head made me think of an anatomical model, with flesh, muscle, and even bone stripped away to reveal the underlying structures always hidden from view. Who would have believed that so much machinery of such bewildering complexity lay beneath the smooth surface of the skin? Like Martha, I wanted to look away, yet I could only stare with fascination at the terrible, unfamiliar sight.

I wasn't so much dispassionate as willfully detached, choosing to see the ravaged head as belonging to no one in particular. I tried to see it as the model it resembled, as a plastic or latex representation, replete with vivid colors, yet having nothing to do with Walt.

Walt's eyes were closed and his ventilator-assisted breathing so shallow as to be barely perceptible. He lay motionless on the cranked-up hospital bed, sedated to the point of unconsciousness. While Michael probed the bloody wound to recover deeply buried gauze, Walt showed no sign of pain. His skin was sallow and waxy, like the thick, whitish tallow covering the vivid red flesh of an uncooked beef roast.

I was at Walt's bedside because he was a friend. A friend brought low by a virulent cancer. He'd been a handsome man. Tall and wide-shouldered, with a big head, beaky nose, and jutting chin. His hair was curly and midnight black, with gray frosting at the temples. He thought fast, talked fast, and laughed loud. He seemed surrounded by a field of energy. This made him fun to be around, because you could absorb some of his energy and feel that you too were lively and clever.

At fifty-six, Walt was twenty years older than I was, and the age gap seemed too wide for us to be close friends. Yet since he'd introduced himself to me at a talk I'd given on the philosophy of science at the medical school five years earlier, we met for lunch several times a year and talked, often for two or even three hours. I was Walt's sounding board and sparring partner, his connection to the world of ideas and principles.

Walt would sip his Manhattan, light up a Kool, and look me straight in the eye, as if challenging me to tell him nothing but the truth. He switched to tea at the end of the meal, but he would go back to smoking. He smoked one cigarette after the other while we talked, drawing the smoke deep into his lungs, then expelling it in a puffy gray cloud. He would watch the cloud drift up, following it with his eyes while he framed his next point.

We always ate at Yen Chin, a large Chinese restaurant near Walt's office, a place where he was known and we could count on not being disturbed. He never invited me to his house, and although I knew he was married, I never met his wife, Martha. Walt and I were intellectual friends, rather than personal ones, each of us an emissary from a different world the other found intriguing.

Walt was a psychiatrist in private practice, yet he rarely mentioned his patients. He was most interested in talking about psychiatric theory, in particular the classification of people as neurotic (then a technical term) or psychotic on the basis of their behavior. He was a supporter of the radical psychiatrist Thomas Szasz, who argued in a famous article,

"The Myth of Mental Illness," that mental illness does not and cannot exist.

The mind, Szasz claimed, isn't the sort of thing that can be ill, and to think that it can is to make a conceptual error, like thinking that a sick joke is a joke that's not feeling well. Szasz believed, in current jargon, that mental illness is a socially constructed category, one that society fills with the various sorts of behavior that it finds objectionable.

"Szasz is defending a half-truth," I said to Walt during our first lunch. "Maybe no more than a sixteenth-truth."

"No, no, no," Walt said, struggling to get the words out fast enough to reject my criticism before I could defend it. "Psychiatrists don't cure patients. Surely you admit that."

"Not necessarily the way other doctors do," I agreed.

"That's not even their aim." Walt threw up his hands, gesturing toward the ceiling and making his jacket fly open. "We treat patients so that we can *control* them." He pointed his finger at my chest, as if holding me responsible.

"Control their illness?" I was puzzled, because I hadn't read Szasz at the time.

"No." Walt barked out the word. "We lock them in the closed wards of hospitals, then stick them in so-called asylums. We drug them and shock them and even operate on their brains." He made a shoving motion with his hands. "We do whatever we need to do to get them out of society and out of our way."

"And that's how you treat your patients?" I asked in a skeptical tone.

"No, I don't," Walt admitted. "Most are people having a hard time who need a little support." He wagged his finger in the air. "And that's my point. Because they feel sad or have trouble making friends, we call them mentally ill."

"So there aren't any crazy people?"

"Not the way psychiatry means it," Walt said. "Look how Soviet psychiatrists lock up people for not accepting communism. If people can't see the truth of Marxism, they must be suffering from disordered thinking. Thus, they're crazy and a threat."

"But that's political repression," I said.

"So?" Walt almost shouted. "It's not political repression when we lock up a woman for believing the president is from Mars? Or for babbling in the grocery store?" He held out his hands, palms open. "Don't you see? We want these people off the street and out of sight. So we call them crazy and turn them over to psychiatrists."

"What about the ones whose neighbor's dog tells them to shoot somebody?" I asked. "Like the Son of Sam?"

"We may have to lock up dangerous people," Walt agreed. "But we don't have to call them mentally ill and try to cure them." He paused. "We need to let people decide how to live their lives, not force them to follow our social blueprint."

We talked about other topics, but mental illness was the one we kept coming back to. Over the years, I got Walt to acknowledge that brain disorders can affect behavior, so people we label schizophrenic might have something wrong with their brains. Yet he clung to the idea that, until proved otherwise, mental illness—which included "neuroses" like homosexuality, narcissism, phobias, drug and alcohol addiction, and extreme shyness—is nothing but a catalogue of behavior society dislikes and wants eliminated. I agreed with him in most cases and, years later, so did the American Psychiatric Association.

One late October, after a lapse of about six months, Walt and I met for lunch at Yen Chin. When I arrived, he was already sitting at our usual table in the dim back corner. I waved to him, but when I walked closer, I was stunned by his appearance. He was stoop-shouldered and so shrunken that his gray pin-striped suit hung on him in loose folds, crumpled and wrinkled like a half-emptied bag of flour. His shirt collar was a couple of sizes too big, as if borrowed from a larger man. His dark, wavy hair was sparser, and the skin of his hands and face was papery and dry. His mouth was shrunken, his lips turning inward, and his nose, always prominent, looked sharper and more beaky.

"What the hell is wrong with you?" I asked. I was shocked beyond the point of politeness.

"Bad news," Walt said. His voice was a rasping whisper. "Get yourself a drink, and I'll tell you about it."

"Just tell me," I said.

Walt held up his hand, and the waiter brought me a neat scotch and a pitcher of water. While I splashed a little water into my glass, Walt stirred his Manhattan with a green swizzle stick. He was the only person I knew who drank concoctions I associated with the sophisticated nightlife of 1950s movies. I noticed, though, that the other element of that sophistication was missing. Walt wasn't smoking. His usual pack of Kools wasn't on the table, and the ashtray beside him was empty.

"So?" I said impatiently.

"About ten years ago, I was diagnosed with throat cancer," Walt said. His voice, despite its raspiness, was undramatic. "It was caught early. An ENT surgeon cut out the tumor and said I wouldn't get any benefit from chemotherapy or radiation."

Walt cleared his throat, then gave a sharp cough. "I spent two days in the hospital, then two weeks at home. Like having your appendix out, that easy." He took a sip of his drink and smiled. "My voice was weak, but my job doesn't require much talking, so I went back to work."

"And you went on smoking?" I at once regretted asking such a judgmental question, but I was still too upset to control my immediate responses.

"I went back to it." Walt's voice was nearly a croak. "I made a deliberate choice to smoke, because I like it. And I knew the risks." His eyes caught mine and he smiled. "I could have stopped, the way any addict can who wants to. You know my views."

What he meant was that if mental illness really doesn't exist, then people are wholly responsible for their actions and can't make excuses based on some supposed aberrant mental condition. Walt smoked, in his opinion, not because an addiction forced him to, but because he decided to smoke.

We had debated the topic often. Szasz testified that kidnapped heiress Patty Hearst had taken part willingly in the bank robbery for which she was being tried, but I argued that months of living under the constant threat of death for disobeying orders had weakened her will. She had acquired the habit of doing whatever her captors told her. Walt, along with Szasz, held that the SLA had convinced her to give up her bourgeois life and become a revolutionary.

"So what's happening to you is recent?" I asked. I took a big swallow of scotch, hoping the alcohol would loosen the knot in my stomach.

"I was fine for years," Walt said. He gave a series of hacking coughs that brought a pained expression to his face. He wiped tears from his cheeks with a folded blue handkerchief. "Checkups every six months, then every year. Nothing turned up until ten months ago."

"You didn't mention anything last time," I said. "That was what . . . about six months ago?"

"Then I only had a catch in my throat when I swallowed," Walt said. "Like food was getting hung up on the way down." He shrugged. "I've had a lot of scares, and I thought that was just another. But I did go to an ENT, and he didn't see anything." Walt sipped his drink and gave a rueful smile. "Trouble was, he didn't use a fiber optic laryngoscope, so he couldn't see far enough down."

"Oh, God," I said, throwing myself back in my chair. "Doesn't any doctor ever get a diagnosis right on the first try? Even one where you just look and see?"

"Not this time." Walt shrugged. "I wanted to believe the ENT, and that helped me fool myself. But a couple of months later, I knew something was definitely wrong. I was having trouble swallowing, and I was so hoarse patients began asking me if I had a cold."

Walt coughed sharply, then cleared his throat. "Allergies, I told them. Some of the time, I even believed it." He smiled and shook his head. "Never underestimate the power of denial. It's not just a river in Egypt."

"Very funny," I said. "Did you go back to the ENT?"

"To another one—Roy Stone." Walt said. "Roy did a complete workup—laryngoscopy, sonogram, then CT scans and biopsies. He found new tumor growth at the old site."

"So you had more surgery?"

"That's affirmative," Walt said. He twisted his head sharply to the side and displayed a scar starting at the hinge of his jaw and disappearing under his collar. It was a puckered line so red it seemed raw. The dark marks of sutures along the sides looked like nail holes.

"And I got chemotherapy this time." Walt put his hand to his throat, bent his head, then swallowed hard. "I finished it this last week."

"You're looking good for somebody who's been through so much," I said. I tried to sound upbeat.

"Yeah, I could tell from the way you greeted me just how great I'm looking," Walt said. He smiled and sat up straight. "But I'm buying time. Since Martha and I don't have kids and she hasn't been a social worker for years, unless you count taking care of me, I need to leave her as much money as I can. So I'm seeing more patients, then moonlighting at the VA."

"Maybe with the surgery plus the chemo, you'll get another ten or twelve years," I said.

"In medicine, anything's possible," Walt said. He swallowed some of his Manhattan, then leaned forward, as if about to tell me a secret. "Problem is, where cancer is concerned, the guys who deal with it don't always know when to quit."

"And you do?" I asked.

"I hope so," Walt said. He cleared his throat with another hacking cough. "I'm not going to put Martha through hell just to get another six weeks of lying in a hospital too doped up to recognize anybody. What I'd like, when things look hopeless, is for somebody to arrange to send me off peacefully." He suddenly smiled. "Euthanasia, as you probably know, is entirely in keeping with my principles as a radical psychiatrist."

"Personal responsibility," I said, nodding.

"I should have asked your permission first," Walt said abruptly. He dropped his gaze and focused for a moment on the empty ashtray. "I hope you won't object."

"Permission for what?"

"I told Martha to call you if she decided it was time for me to go and needed somebody to back her up." His eyes caught mine and held them. "You know how hospitals work, and you know what I believe. More than anybody else does, really."

"I should," I said. "After years of listening to you rant about dignity and autonomy." I laughed, but I felt a deep chill.

"Great," Walt said. His voice cracked, and he was forced to clear his throat repeatedly and drink some water. "Now let's get back to basics and talk about freedom of the will."

"I'm sure you're not supposed to be having that Manhattan," I said. "No ENT or oncologist would allow a cancer patient to drink alcohol."

"That's where we can start the discussion," Walt said. His voice turned into a croak before he could finish the sentence. "About my freedom to disobey my doctor's orders."

I thought about Walt off and on during the next four months. More time than usual had elapsed between our lunches, and I considered calling and asking how he was. But I didn't feel comfortable doing it, because we didn't have that sort of relationship.

We were friends only in a prescribed, intellectual way, the way you might be friends with a teacher from your past whom you meet for coffee occasionally. I would have felt intrusive, almost prying, had I called Walt.

Yet I may be excusing myself too easily. Someone less cowardly would have phoned without a second thought. I was afraid to ask Walt how he was, because I felt I knew the answer. Cancers of the head and neck, when caught only at a later stage, almost always have a fatal outcome.

Then, finally, one morning Martha called me at home. She said she needed to talk to me and asked me to meet her at three o'clock in the twentieth-floor solarium of Central Hospital's Cancer Care Center. It was late April, almost May, and the weather was cold but sunny.

"How's Walt doing?" I asked her on the phone.

"Holding on," Martha said. She hesitated, then said, "I'll tell you more when we get together."

I had never seen Martha, so I wasn't sure I would recognize her. "I'll be wearing a charcoal gray jacket," she told me. "I'll try to get a seat on the left as you come through the door."

Her precautions for avoiding confusion turned out to be unnecessary, because she was the only person in the solarium. Even so, she was sitting where she said she would be. When she saw me, she got up and took a few steps toward me.

"Ron? I'm so glad you came," she said. Her voice was warm, and she forced a small smile. She held out her hand. "I hope you didn't have trouble finding this place."

"None at all," I said.

Martha was short and dumpy, considerably overweight, with gray hair cut in a short bob that made her face look very round. She seemed housewifely in an old-fashioned way that surprised me. Walt was tall and good-looking and projected immense vitality, so, quite unconsciously, I had expected his wife to be a female version of him.

"How long has Walt been here?" I asked.

"A week tomorrow," Martha said. "We checked him in last Tuesday."

We sat on a blonde wood bench beside a long, white stucco planter filled with a variety of caladiums with broad, mottled leaves. Ficus trees in large pots were scattered about on the red tile floor. The solarium's roof and walls were glass panels set in aluminum frames. The sun was going down, and the light coming through the glass at a low angle caused the plants to cast long, stark shadows. Shadows from the caladiums looked like bundles of sharply pointed knives.

"It's already too long," Martha said. Her eyes were red-rimmed and puffy, and up close I could read the sadness in her face. Yet her voice was firm, and I sensed her steadiness. She was deeply upset but holding herself together by an act of will. Her dowdiness, I suspected, disguised a tough, unyielding character. "That's why I need your help."

"I'll do whatever I can," I said.

We were up so high that we looked down on the dozens of shorter buildings surrounding the hospital tower. Many had patios or roof gardens with flower boxes, but the weather was still too cold for much of anything to be growing. The hospital plants were flourishing only because the glass walls provided them with a protective environment.

"He's had enough," Martha said. She looked squarely at me and spoke in a flat, resolute tone.

"What do you mean?" I asked.

"The chemo didn't kill the cancer." Martha shook her head. "Dr. Stone said it hadn't responded well and was very aggressive. It spread to Walt's tongue and to his upper and lower jaws." She laid a hand flat against the right side of her face.

"This was when?" I asked.

"The chemo was over in February," Martha said. "They knew at once it had failed." Her voice faltered for the first time, and she cleared her throat. I was reminded of Walt's harsh cough and raspy voice. "Then there was a disagreement."

"About what?"

"The oncologist, Dr. Alverz, wanted Walt to have radiation, which wouldn't cure him, but might slow down the cancer." Martha's voice was firm again. "Dr. Stone, the surgeon, said he might be able to remove all the cancer, which could mean a cure. But he couldn't operate after radiation. Something about the tissues not healing properly."

"And Walt opted to go with the surgery," I said. "That's what I'd expect him to do."

"He doesn't do things by halves," Martha said, with a hint of pride. "That's why I tell him I don't want him to gamble. It would always be double or nothing."

"When was he operated on? Wednesday?"

"No." Martha shook her head. "They ran tests Tuesday and Wednesday, then took him for surgery on Thursday." She closed her eyes tight for a moment, pressing her lips together. Finally, she said in a fainter voice, "The surgery took twelve hours. Twice the length Dr. Stone said to expect."

The lengthy operation suggested Dr. Stone had spent a lot of time identifying cancerous tissue and eliminating it. But if the disease had infiltrated Walt's mouth and spread to the bones of his jaw, Dr. Stone must have realized at some point that cutting out all the diseased tissue was impossible.

"Dr. Stone warned him," Martha said. "He told Walt he might have to remove some or all of his tongue." Her voice faltered, becoming almost inaudible. "And he might have to take out some jawbone."

Martha opened her purse and pulled out a white handkerchief with tiny pink flowers embroidered on one corner. She blew her nose gently, then put the handkerchief away.

"Walt knew the risks and still consented?"

"Yes, he did." Martha's tone was normal again. "Of course neither of us imagined things could turn out so badly." She patted the side of her face. "Dr. Stone took all of Walt's tongue, and instead of pieces of bone, he removed the whole right side of Walt's upper and lower jaws."

Tears welled up in her eyes. "Dr. Stone said everything he cut out was riddled with cancer." She paused, then said, her voice rising into a wail, "Now Walt's completely mutilated."

"But he's alive," I said.

"Part of me is glad," Martha said, nodding. "I want him to be alive. But not like he is. And not with what's ahead of him." Her eyes strayed to the long spindly shadows cast by the branches of a ficus tree. "I'm sorry he came out of the anesthesia."

"What about the cancer?"

"All for nothing." Martha sounded suddenly tired, almost defeated. "Dr. Stone's stopped talking about a cure, but he wants to operate again. He wants to remove some bone at the base of Walt's skull and replace it with a titanium plate. That'll give Walt some extra time." She squeezed her eyes tightly shut and lowered her head. "Then when Walt is better, he says, they can start reconstructive surgery, using bone from Walt's hips to build up his jaws."

I sat still, saying nothing. Not knowing what to say. I couldn't tell Martha that Walt was going to be all right or that the surgery would be worth it. I didn't believe either. I thought Walt was going to die soon, whether he had more surgery or not.

"He's had enough," Martha said. She looked up at me, and her voice was strong again. She'd made the same statement earlier, and I hadn't asked what she meant. Now I didn't need to.

"Dr. Stone can't operate until Walt regains consciousness and gets stronger," I said. "You can refuse to authorize more surgery."

"Go with me to talk to Dr. Stone," Martha said. "I want him to stop treating Walt."

"You mean you want to change doctors?"

"No." She gazed over the bare, gray rooftops of the city. "I mean I don't want anybody to treat him."

"I understand." I felt neither surprised nor horrified, but my heart seemed to thicken with a deep sadness. "I don't know if Dr. Stone will agree."

"Then we'll have to convince him." Martha turned to look at me. In the fading light, her round, puffy face was as wrinkled as a walnut, but it also looked strangely youthful, as if she were a young girl aged suddenly by sorrow. "You've got to tell him about Walt. Walt said you know what he believes."

Dr. Roy Stone met us an hour later in one of the hospital's small, anonymous conference rooms. He was tall, middle-aged, and slightly stooped. He was bald, except for a fringe of gray hair, which was clipped so close that it was more fuzz than stubble. Pouches of dark skin under his eyes and heavy lines running from his nose to his mouth made him look exhausted.

Dr. Stone and I shook hands when Martha introduced us. He then gestured toward a small table and four chairs, the only furniture in the room. We sat down, our knees almost touching.

"I checked on him before coming here," Dr. Stone said. He was looking at Martha, but I found it odd that he didn't use Walt's name. "His blood pressure is lower than I'd like, and we're giving him drugs to bring it up. But he's stable and should be out of danger in a couple of days."

"He's had enough." Martha said it to Dr. Stone the way she had said it to me: a statement of fact that no reasonable person could contradict.

"I can imagine how upsetting this is." Dr. Stone spoke gently, considerately. "But before long, Walt will be strong enough for us to remove the affected tissue I mentioned."

"Cut out part of his skull," Martha said. "I remember." She looked miserable, but her voice stayed strong. "Maybe he'll get a brain infection from that. But if he doesn't, you'll start the reconstructive surgery."

"When he's recovered enough, yes," Dr. Stone said. He paused, frowning. "And I can't deny that things can go wrong. Infection is possible, and so is a bad reaction to the anesthesia. So is a stroke during surgery."

"Walt wouldn't want to go on with all this." Martha gave Dr. Stone an unflinching look. "If he was conscious, he'd tell you to stop treating him and let nature take its course."

"With due respect," Dr. Stone said, "I'm not sure you can read your husband's mind." He locked his fingers together on the table, turning his hands into a single large fist. "He struck me as a fighter."

Martha remained silent and didn't look at me. But I sensed that she was ready for me to join the conversation.

"Dr. Stone," I said, "if you were in Walt's situation, would you want more surgery when you'd already had half your face removed and still had cancer?"

"Maybe." He sounded positive, upbeat. "But I admit I might not." He gave me a tight, professional smile. "The decision is personal, and we can wait a couple of days until Walt is better and ask him what he thinks."

"Walt wouldn't want to get even that far," I said. "As Martha said, he'd want you to quit treating him right now."

"He'll die if I stop," Dr. Stone looked at Martha and held her gaze for a moment. It was an open challenge.

"That's what he wants." Martha nodded her head to emphasize her statement. "He wants to die."

"Walt would consider that the best of the available outcomes," I said. "He and I have talked about freedom and responsibility often for the last five years. I can't read his mind either, but I know his thoughts so well that I can predict what he would say. He's a radical psychiatrist."

"I don't know what that means." Dr. Stone unclasped his hands and sat back in his chair. The discussion wasn't taking a course he was familiar with. He seemed both surprised and puzzled, yet willing to listen.

I explained Walt's principles.

"He believes individuals are responsible for their actions," I said. I spoke slowly, earnestly, wanting to state Walt's views accurately. "People must be permitted to run their own lives, even if society or other people disagree with them."

"And I'm guessing that means ending their lives." Dr. Stone puts his hands flat on the table. "If they want to. And for whatever reason seems to them a good one."

"You've got it," I said, nodding. "He believes in rational suicide, and he believes in euthanasia."

"But does he think other people can make the choice for him?" Dr. Stone asked.

"Only in special circumstances," I said. "Euthanasia can be self-directed and self-administered, when possible." I glanced at Martha and saw she was sitting quietly, concentrating on what I was saying. "The last time I saw Walt, he told me he didn't want to be kept alive if the quality of his life was going to be poor. If he was unconscious or incompetent to make a decision, he left it up to Martha to decide when to end his life."

"He told me exactly the same thing." Dr. Stone sighed, then rubbed his forehead and massaged his eyes with his thumb and index finger. "I told him not to be so pessimistic."

"He was being realistic," I said. "Walt could face facts, even when he didn't like what he saw. He was an intellectual, but he wasn't a wooly-headed dreamer."

"I take your point." Dr. Stone nodded at me. "But surgery patients, particularly cancer patients, need to be encouraged. Even though Walt is a physician himself, talking about euthanasia right before surgery isn't reassuring."

"He told me he'd like it best if you could give him an overdose of some painkiller," Martha said.

"He told me that too," Dr. Stone said. "But I can't do it, even if I think it's justifiable. It's not legal, and I don't want to be charged with homicide or lose my license."

"He didn't expect to be in such a terrible state," Martha said. "He didn't expect to be so…so maimed."

Dr. Stone screwed his face into a frown, as if a sudden, sharp pain had passed through him. "I didn't expect it either," he said in an almost plaintive tone.

Dr. Stone, I realized, wasn't a surgical cowboy, somebody ready to do anything to his patient to gratify his sense of power. He saw that he'd gotten in over his head with Walt by trying to do too much. He'd wanted to cure Walt, not merely give him a few weeks of extra time. His patient was now worse off than if he'd done nothing—a bitter experience for any physician.

The three of us sat still. We didn't look at one another, and no one fidgeted. Martha's head was bowed. Despite her silence, she seemed to radiate grief, and I felt an impulse to draw away from her. I wished I were anyplace else, then I felt guilty for wishing it.

"I don't know what I can do," Dr. Stone said. "Not really." His voice was soft, and he didn't seem to be speaking to either of us. His elbows were propped on the table, and he rested his head on the tips of his splayed fingers. Like Martha, he was staring at the polished top of the table. But, instead of grieving, he appeared to be deliberating, ransacking his mind for an acceptable course of action.

"Are you sure he wouldn't want to go on?" Dr. Stone folded his arms across his chest and looked at Martha. "Are you absolutely certain?"

"Without a shadow of a doubt." Martha raised her head and looked into Dr. Stone's eyes. "It's time to let him go." She turned to me. "Isn't it? Isn't that what he would want?"

"He would." My mouth was dry, and my tongue felt sticky. I cleared my throat and swallowed. "If Walt was able to give himself a lethal dose of morphine, that's what he would do."

"Right." Dr. Stone spoke absently, as if distracted by a puzzle he was trying to solve in his head. Then he abruptly stood up and said, "Let's go visit him."

Michael was in the process of changing Walt's dressings when the three of us arrived in the room. Michael looked at Martha and me, and

I saw from the look that flickered across his face that he was about to object to our being there. Then he noticed Dr. Stone, nodded to us, and turned his attention back to tending to Walt's surgical wounds.

"Don't let us interrupt you, Michael," Dr. Stone said. He held up a hand, as if warding off an objection. "But let me take over after you remove the packing." He paused, then, almost mumbling, said. "I want to examine the wounds."

"You're not going to have long," Michael said. "When the packing is out, he loses blood fast. It's a struggle to change the dressings and keep him from bleeding out."

"I'll only be a minute," Dr. Stone said.

Michael picked up the plastic handpiece of the suction device, flipped the tube to get a kink out of it, then began to suck up the blood pooling on the floor of what remained of Walt's mouth. The slurping noise was like that made by a soda straw at the bottom of a glass. It was both familiar and alien, and it made me feel a little sick to my stomach.

I looked at Martha and saw that she was stone-faced. Her heavy chin was set, and she watched every move Michael made. Yet her hands were wrapped so tightly around the metal railing at the foot of the bed that her fingers looked white.

Dr. Stone pulled on a pair of thin latex gloves from the box on the tray behind the bed. "I'll take over," he told Michael. He put his hand at the top of the suction. "Please wait outside, and I'll call you in a few minutes."

"You want to do the dressings yourself?" Michael sounded doubtful. "Let me know when you finish your exam, and I'll repack the wounds. I've got the supplies laid out."

"I'll do what needs to be done," Dr. Stone said.

Michael left the room, and Dr. Stone suctioned away more of the accumulating blood. He hung the handpiece in place on the equipment console, then switched off the alarms on the heart monitor and the mechanical ventilator. The ventilator kept working, hissing slightly as it pushed air into Walt's lungs through the endotracheal tube in his throat.

Blood was draining from Walt's wounds as rapidly as Michael had warned. Dr. Stone vacuumed it away with the tip of the wand, but it was obvious that Walt's mouth would soon be full again. Dr. Stone replaced the suction handpiece in its holder. He then unwrapped a blue surgical towel from its sterile bag and draped it over the ruined right half of Walt's face.

"Martha," he said, turning toward her. "Why don't you come up here so you can be closer to your husband."

Martha nodded, but she hesitated. I realized that she was afraid of collapsing if she turned loose of the bed frame. I took her arm by the elbow. Together, moving awkwardly, we walked to the head of the bed. Martha, now closer to Walt, seemed to gain strength. Dr. Stone stepped aside to give us more room.

Strangely, with the destroyed part of his face covered, Walt looked more familiar. Despite the tube coming out of his mouth, he was recognizable as a friend, and I no longer saw him as an anatomical exhibit. Yet, up close, the waxy, yellowish-white color of his skin looked even more like tallow, and the musty odor of urine, plus the metallic smell of fresh blood, hung in the air around him. He was Walt; yet, simultaneously, he was not Walt. The resemblance was close but not exact.

Martha put her hand against Walt's left cheek, then smoothed back his lank, greasy hair. She leaned forward and kissed him on the forehead. "Goodnight, my prince," she said in a low voice.

I noticed a thin line of blood running from under the towel covering Walt's face. The blood was trickling down Walt's neck and staining the white sheet covering him. The small spot was quickly becoming a large, irregular blotch.

"Wait outside now, both of you." Dr. Stone's tone was brusque and peremptory. Then he added, "Tell the nurse to stay out there until I call for him."

Martha looked at Dr. Stone for a long moment, then without a word she turned. I tried to take her arm, but she ignored me. I held the door for her. Glancing backward before the door closed, I saw Dr. Stone with the suction. He was again vacuuming away the blood leaking from Walt's terrible wounds.

Michael was standing at the nursing station doing paperwork, but he looked up as we came into the waiting area.

"Did he want me?" Michael asked.

"No," I said. "He said for you to stay out here until he calls you."

Michael put down his ballpoint pen and frowned. He appeared puzzled, and for a moment it looked like he might go into the room, despite Dr. Stone's instructions. Then he picked up his pen and went back to his paperwork.

Martha and I sat on the gray upholstered chairs across from the nursing desk. Neither of us spoke. She put her black purse in her lap and folded her hands on top. She closed her eyes and blinked, but I couldn't tell whether she was trying to keep from crying. She seemed tired and resigned, beyond the point of obvious upset. She looked like a middle-aged traveler who had missed her bus and knows she must wait several hours for the next one.

I looked at the white face of the clock above the nursing desk and saw that it was past five. I seemed to lose track of time after that. I thought about the papers I had to grade and the article I needed to revise for a journal. I felt a stab of guilt but shrugged it off. Walt would have considered it natural and healthy to be thinking about getting on with life.

I wondered if I ought to ask Martha if she wanted coffee. I would be glad of an excuse to get out of the waiting room. Martha said nothing, but her misery was palpable. My feelings about Walt seemed so weak in comparison that I was more aware of her sorrow than of my own sense of impending loss.

The door to the SICU room suddenly swung open and Dr. Stone came into the waiting area. His face was pale, almost ashen, and the lines between his nose and the corners of his mouth seemed even deeper. His eyes were opened wide, almost staring, and the sleeve of his white lab coat was splattered with drops of blood. Martha and I stood up and took a couple of steps toward him.

"Martha," he said, putting a hand on her shoulder. "Walt has passed away." Dr. Stone's voice was unsteady, and he paused, then cleared his throat. "His blood pressure was unstable, and while I was changing his dressings, it dropped very low. Probably because of bleeding. His heart eventually went into an arrhythmia that I couldn't reverse." He shook his head. "It happened very quickly."

"I understand," Martha said. She nodded, and tears began to trickle down her face. "I understand."

"He slipped away peacefully." Dr. Stone took his hand away from Martha's shoulder, then didn't seem to know what to do with it. He stuck it in the pocket of his white coat.

"I'm glad," Martha said. "He's been through enough. Too much, really." Her voice was thick, and she wiped away her tears with the small handkerchief she had used earlier. "I want to thank you. You did your best for him, and I'm grateful."

"Do you want to see him?" Dr. Stone asked. "I can arrange for you to spend some time alone with him."

"I don't need to," Martha said. "I've already said goodbye." She folded the handkerchief and wiped her nose, then she turned to me. "Will you have dinner with me?"

"Of course." I had been about to say I wasn't hungry and couldn't imagine eating, but I caught myself.

"Would it be okay if we go where you and Walt always had lunch?" Martha pressed her lips to form a slight smile, but more tears filled her eyes. "That seems fitting to me."

"To me too," I said. "And for the first time in my life, I'm going to order a Manhattan."

The saddest of truths is that death and suffering come to us all. And not only to us as individuals, but to our wives or husbands, our parents, our friends, and even our children. This truth is what Virgil calls *lacrimae rerum*, the tears that are in things.

We hope that when death comes it will be swift and will allow us to depart without prolonged suffering, our dignity intact. We hope also that it will not exact a cruel price from our family and friends, forcing them to suffer as we linger for days or weeks in a hopeless condition, waiting for release.

THE POWER OF MEDICINE

Not so long ago, people could count on death bringing a quick end to suffering. If an old man, frail and weak, were hospitalized with bleeding into the brain and then developed pneumonia, he would most likely die within hours. Medicine could provide patients comfort and encouragement, but in the face of devastating diseases, it could do little to stave off death.

The advent of antibiotics in the 1940s marked a decisive change. Now the old man can't be expected to die of pneumonia (the traditional "old-man's friend" of medicine). If his infection is cured, his doctors may decide he could benefit from surgery to tie off the leaking vessel and remove the pooling blood pressing on his brain.

The surgery has risks. The procedure itself may damage his brain, or a drop in his blood pressure may trigger a devastating stroke, or the swelling of his brain may destroy brain tissue. The old man may never regain consciousness, yet he may remain alive for an indefinite time. Possibly years, with proper care.

Medicine has acquired such power that even when it cannot restore health, its drugs, machines, surgical procedures, and disease-management techniques can often postpone death beyond the point where most thoughtful people want to remain alive. Death postponed may become nothing but dying prolonged.

Only a few decades ago, when an injury or disease was serious, instructing a doctor to "do everything you can to keep her alive" meant little. Not much could be done. We've learned to avoid such instructions, because now so much is possible.

DECISIONS

When we exercise our autonomy, we are typically deciding how we want to live. Do we want to risk our necks attempting to climb K2 or sit on the sofa watching reruns of *Law and Order*? Yet we are also exercising our autonomy when we decide whether to discontinue therapies that are keeping us alive.

Medicine and the law have both recognized the right of patients to discontinue even life-sustaining treatments. They have also recognized the legitimacy of allowing a family member or a close friend to make that decision for someone not in fit shape to make it for herself. (As I pointed out in the Prologue, the Supreme Court decision in the Quinlan case marked the beginning.)

Most of us have a firm idea about whether we would want to remain on life support if we had no realistic hope of breathing on our own or regaining consciousness. We know what we would choose if we had little chance of prolonging our lives even after undergoing multiple surgeries. Most of us would favor allowing our lives to come to a close as quickly and painlessly as possible.

Usually we also know what our friends and family members think about being kept on life support or subjected to painful treatments with only a remote chance of success. Even if we haven't discussed the question with them, we've picked up on their attitudes. ("I don't know what Joan's family was thinking, leaving her hooked up like that for an entire week. They knew the first night that she had irreversible brain damage and wouldn't be able to breathe on her own.") We also extrapolate from what we know about people's personalities and principles. Martha Post had no doubt what Walt would want done, had he been able to announce his decision. She knew his views about independence and personal responsibility. Decisions to end treatments aren't hard to make because we doubt what somebody would want done. They're hard because we have to live with our doubts. What if we are mistaken?

Suppose Sam died after we ordered his treatment ended. We gave the order because we believed he wouldn't want to be kept alive as a severely brain-damaged person who would never regain consciousness and would spend the remainder of his days fed through a gastric tube and hooked up to a ventilator. Then, a week later, we find a document—written and signed by Sam—instructing his doctor to keep him alive as long as possible, regardless of his condition or prospects.

We made a mistake about what Sam wanted, but we didn't commit a serious wrong. Our decision was based on the notion that no reasonable person would want to continue to exist in Sam's condition. Our judgment was correct, given that assumption. The judgment that no reasonable person would want to continue to exist if he needed an appendectomy and a two-day hospital stay would be incorrect.

Sam lost his life, but he was deprived of nothing that reflective people consider makes life worth living. Indeed, keeping Sam on life support without knowing his wishes would be a more serious wrong than terminating his life support.

HOLDING BACK

Doctors, for the most part, are no more eager to impose treatments likely to be futile than most patients or their families are to want them. Every day at every hospital in the country, ventilators and monitors are removed, antibiotics are stopped, blood-pressure medications are withdrawn, and people are allowed to die. The door is opened to the disease that will carry them away.

Just like in the old days.

"Do not resuscitate" (DNR) orders are often written in patients' charts, at the request of patients and families. If a patient with a DNR order goes into cardiac arrest or septic shock, no action is taken to pull him or her back from the edge. Similarly, although gastric feeding tubes are removed, families and the nursing staff pray that some quick-acting infection will carry off the patient, sparing them the stress and sadness of watching a lingering death. Patients are given large doses of morphine to reduce their pain. The drug also slows respiration, so an amount effective in controlling pain can, in a weakened patient, hasten death. Blood loss from seeping surgical wounds is a less common cause of death, but it too is quick and painless.

The wise person, the stoic philosopher Seneca said, doesn't ask how long can I live, but when is the right time to die.

SEVEN

Not More Equal

"Ron?" The woman's voice was tentative, then confident. "This is Joan Marvin. I'm sorry to bother you so late in the evening, but we need an ethics consult."

Joan Marvin was professor of surgery and chief of thoracic transplant surgery at Midwest School of Medicine. We had met several years earlier at a reception promoting organ donation, where we chatted over glasses of white wine. Each of us was speaking that evening, though, so we were too distracted to have a real conversation.

Joan, in her talk, defended altruistic donation, while I argued for selling organs in a regulated market. Our exchanges during the discussion period were courteous but sharp, so I was surprised when, at the end of the reception, she asked me to give my same talk to the residents in her transplant program.

"They need to hear some fresh ideas." Joan's smile was almost malicious. She seemed pleased with the idea of showing the residents that she wasn't as predictable as they might think.

This openness impressed me, and Joan and I had stayed in touch since then. We exchanged pleasantries when we happened to meet, and she once sent me an e-mail asking my advice on developing a policy on living lung-segment donors. I wasn't completely surprised by her phone call.

"I was only reading a novel," I said, brushing aside her apology. "It's not even eleven. When do you need the consult?"

"Tonight," Joan said. "Can you come to the hospital?"

"Right now?" What she'd said didn't register.

"If you can." She remained polite, but her voice conveyed a sense of urgency. "Our patient's condition is unstable, and we need to decide whether to list him for a heart. If we do, he goes on as Status One."

"I'll leave right now," I said. Status One meant the patient was being kept alive by intensive means and might crash and die at any moment. "Give me half an hour."

A request for an immediate consultation was rare. Sometimes I was asked about the ethical legitimacy of using an experimental surgical procedure or drug to try to help a patient with no other options. Such cases were urgent, but, even so, I usually had a day or so to make a recommendation. I didn't like the idea of giving advice on the spot, without being able to take time to ruminate about the relevant facts and principles.

I parked in the hospital garage and walked under the blue-white halogen light to a side entrance of the main building. The security guard, a heavy-set African-American woman in a blue uniform, was sitting at a desk both looking at a newspaper and watching a small TV. The large, high-ceilinged lobby beyond was dim, shadowy, and silent, like a grand ballroom after the party is over, the lights have been lowered, and everyone has gone home.

"Dr. Marvin said she'd tell you to expect me," I said.

"Picture ID." The guard spoke the words in a low, muttering voice, conducting a ritual so familiar to her that she didn't see the need to speak clearly or even audibly.

She took the university card I held out and studied it, glancing from the photograph to my face. I was wearing a tie, but I wasn't in the picture. What if she didn't see the resemblance?

This added to my anxiety.

Joan had the final call on whether her patient was to be listed for a transplant. But I'd be offering an opinion. It might sway her decision, so I couldn't escape all responsibility. My eyes burned from the dry air in the building, and I experienced the first low, dull throbs of a headache.

The guard handed back my ID and returned to her newspaper and television. My uneasiness didn't disappear, though, and as I walked through the shadowy lobby toward the bank of high-speed elevators, I felt I was entering into a peculiar netherworld, a grim realm without familiar landmarks.

The fifteenth floor was deserted and dim, the only sounds the low, persistent hum of laboratory freezers and refrigerators. Now and then one would shut off with an abrupt thump, followed by a metallic shudder. Just as abruptly, some other would click on and whirr into mechanical life. The noises filling the corridor muffled my footsteps.

Light spilling from Joan's open door formed a bright fan-shaped patch on the polished terrazzo floor. She was writing at her desk, her back turned toward me. I tapped lightly before stepping inside.

"You made good time." Joan flashed me a brief smile, and we shook hands. "I wouldn't have dragged you out so late on a Saturday night if we didn't need your help."

Joan was short yet charged with energy. She had a long, narrow nose, pencil-thin eyebrows, and snapping dark eyes. The oval lenses of her wire-framed glasses focused her gaze, making it more intense. Her auburn hair was sprinkled with gray, and under her white lab coat she wore a black skirt with a pale peach blouse. Without the white coat, she would have been dressed for a stylish party.

"His name is Rob Dwyer, and he's in the CICU," Joan said. "Let's go down and see him, then we'll meet with the others and talk about what to do. You know Tom Simon, but maybe you haven't met Bruce Mazer and Rachel Hite."

We took the elevator to the tenth floor and walked at a fast clip toward the CICU, the coronary intensive-care unit. The hall was brighter, with more signs of activity. Residents and nurses were gathered around the nursing stations, filling out forms and chatting in low voices. Several room doors were open, and despite the late hour, patients were sitting up in bed watching television. Joan slowed the pace so she could brief me.

"Robert Dwyer is a forty-six-year-old Caucasian with dilated cardiomyopathy," she said. "His heart is so large that it just about fills his chest, but his cardiac output is only fifteen percent of normal. He's edematous, has fluid in his lungs, and is on the verge of kidney failure."

"A transplant could reverse all that?" I asked.

"Probably," Joan said. "He's a good candidate medically. But that's why we need a consult."

"What do you mean?"

She stopped walking and turned toward me. She peered at me closely, the lenses of her glasses making her eyes look even larger. She seemed to be considering how to phrase what she was going to say. She then gave a small shrug, deciding it didn't matter.

"Rob Dwyer is a prison inmate." Joan's voice was husky and low, as if telling me a secret she didn't want others to hear. "He's a convicted murderer."

I felt a shock run though me.

"Who did he kill?" The question was automatic.

"His girlfriend Kim—I don't remember her last name—and her six-year-old daughter," Joan said. "Kim moved out to get away from Dwyer, but he tracked her down and shot them both. The relationship was abusive." Joan leaned toward me, her voice still low. "I don't know any of the details, but he got a life sentence. He's been in prison for ten years already."

"Wow," I said, shaking my head.

A coldness spread through me. A crime of passion I could understand. Or at least I could imagine it. But killing a small child? What kind of person could do that? The irony was too blatant for fiction—a child killer needs a heart.

"I wasn't going to tell you quite yet," Joan said. "But I decided I didn't want you to be shocked when you see him, because he's under guard and shackled." She pursed her lips. "He's so sick that he doesn't need chaining up, but one of the wardens told me they've got to follow their protocol."

I realized suddenly while Joan was talking why she wanted an ethics consult. Rob Dwyer wasn't just any forty-six-year-old man who might be helped by a heart transplant. The question she was going to ask me was, should a convicted killer be given a new heart? A heart that could go to somebody who wasn't a killer.

"Have you mentioned a transplant to him?" I asked.

"Only as an abstract possibility," Joan said. She started down the hall again, reverting to her rapid pace. "But we can talk later."

The rooms of the CICU were arranged in an open rectangle around a nursing station. The woman sitting in a chair beside the door to Rob Dwyer's room had short reddish hair and a round, acne-scarred face. She wore a baggy maroon jacket and shapeless black trousers. I put her down as a Dwyer relative, until I saw the badge pinned to her breast pocket. I then noticed she was carrying a pistol in a side holster.

Despite Joan's briefing, the presence of an armed guard in a hospital was incongruous and disturbing. It made me think of the *Godfather* movies and mobsters carrying out a hit in a place committed to healing.

Joan told the guard my name. She then picked up a clipboard from the counter and wrote on the top sheet. "I've called him in as a consultant."

"He's got to register," the guard said. "That makes you responsible for him."

Joan handed me the clipboard. I wrote the time and signed on the line below her signature. My hands shook, turning my name into an illegible scrawl. I felt uneasy, as if I were about to witness something repugnant. I put down the clipboard, then followed Joan into the room.

Rob Dwyer was thin and scrawny, with a long face, sunken cheeks, and sparse blonde hair. His blue eyes were washed out and dull, his eyelids red-rimmed with gummy, stuck-together lashes. His gaze flickered in our direction, but he gave no sign of recognizing Joan.

"Hello, Rob." Joan's tone was friendly, but brisk. She picked up his right hand, and I saw the handcuff. It was connected by a chain to the one around his left wrist. Light glinted off the chain's chrome finish as the links shifted positions.

On the inside of Dwyer's right forearm, several inches above the handcuff, a coiled snake was outlined in blurred blue ink. The figure was professionally done, but an amateur had applied red ink to sketch in a darting, forked tongue and a single glittering eye. Probably added in prison, I decided.

"How are you feeling this evening?" Joan glanced at the plastic bag of clear liquid hanging from the IV stand, then at the two collecting bags at the side of the bed. One was for urine, the second for the murky fluid draining from Dwyer's swollen abdomen.

"I'm still here, but just barely." Dwyer's voice was hoarse and heavy and so weak it was hard to make out the words. A plastic oxygen line was clipped to his nostrils, and he paused to draw in a breath. "Too mean to die, I guess." He paused, then said again, "Just too mean."

A pale blue blanket was pulled up to his neck, but his right foot stuck out. Both his foot and his ankle were so puffed up that his flesh looked like soft, pale putty. The swelling was the result of the edema Joan had mentioned. Dwyer's heart wasn't pumping effectively, so the fluid that usually would be excreted was building up in his tissues. Breathing was hard for him because his lungs were waterlogged. Despite the edema, a metal band encircled Dwyer's ankle. The shackle was so tight that it indented his puffy flesh.

"This is Dr. Munson." Joan raised her voice to get Dwyer's attention. "He's going to help us decide about your treatment."

I smiled slightly and nodded to Dwyer, but I didn't speak. Also, unlike Joan, I made no effort to touch him or shake his hand. I stood at the end of the bed looking down on him.

"I'm sorry you're sick," I said.

I felt dishonest. Did I care that Dwyer was sick? I'd never spoken to a murderer before, never even seen one. I found being in the small room with Dwyer disagreeable and creepy, as if he were a sick rattlesnake shut up in a cage. I didn't fear him, but I was repelled by him.

"I'm sorry too," Dwyer mumbled. "I need some help."

He fixed his eyes on me, as if pleading, but he said nothing more. His skin was scaly and dry, the color of a dingy white shirt. His lips and nose were tinged with blue, and his breathing was rapid, almost a pant. Despite the tube clipped to his nostrils, he was laboring to take in enough oxygen.

"Dr. Marvin mentioned a heart transplant to you," I said. "How do you feel about the possibility?"

"Scares the crap out of me." Dwyer paused to breathe. "But I feel good about it too." His lips twisted into a smile, and the tip of his tongue darted out for an instant. He shifted uncomfortably in his bed, making the chains around his ankles clink. Softly, his voice catching in his throat, he said, "I don't want to die."

Do you think Kim and her little girl wanted to die?

The question forced itself into my mind from somewhere outside my awareness, perhaps from my unconscious mind. But if I asked the question, I would be violating my obligation to make an unbiased assessment. This would also make my opinion worthless to Joan, who needed some help.

"You remember I explained you'd have to take drugs to keep your body from rejecting the heart," Joan said. "And you'd need them for the rest of your life."

"Whatever it takes, I'll do it." Dwyer's voice was slow and guttural, but strong with conviction. "It's got to be better than this living hell." He fixed his eyes on Joan, then shifted his gaze to me. "Thing is, and this disturbs me a lot..." He paused a long time, then said, "Somebody's going to have to die before I can get me a heart."

I was surprised. I hadn't expected that to occur to Dwyer, much less distress him. But I was also suspicious. Just how sincere was he?

"What bothers you about it?" I asked.

"It seems a shame for somebody to die and for me to get the benefit." Dwyer shrugged feebly. He looked sad, then his face brightened and he smiled. "But if he dies and I don't get the heart, it'll just rot in the ground."

"Or it will go to somebody else who needs it," I said.

Dwyer frowned, looking shocked. I was instantly sorry for what I'd said. Then, just as suddenly, I changed my mind and wasn't sorry at all. Were Dwyer's feelings genuine or was he only trying to con me? I couldn't decide.

But did Dwyer's sincerity even matter? Being sorry about the death of a donor wasn't a requirement for qualifying as an organ recipient. It wouldn't be fair to require Dwyer to meet a standard not demanded of other people.

"Oh, I get what you mean." Dwyer slowly blinked, then let his eyes close. He opened them again, and when he looked at me, he seemed disappointed. "I hadn't thought about that."

"What work did you do on the outside?" I took a few steps closer so I could hear him better.

"Insurance." Dwyer gave a sharp laugh that ended as a gasp for breath. He was then quiet for a moment, as if sorting through his memories. "I was an agent for Life-Star Assurance." He drew in oxygen with sharp snort. "And I was a goddamned good one. I guarantee you I was."

Dwyer's sudden swagger hinted at the way he used to be. Ten years younger, physically powerful, and pumped up with self-importance, he would feel that whatever he wanted, he ought to have. He would strut, bluster, and threaten, and he would tell himself that no bitch was going to run out on him. That's why he'd hunted Kim down and shot her. He'd killed her out of rage to satisfy his inflated sense of himself.

But why did he kill her daughter?

Because she could testify against him? Or was his mind so clouded by blind fury that he'd killed her only because she was unlucky enough to be there? Maybe the blurred tattoo on his forearm revealed a metaphorical truth, and he'd been like a poisonous snake: cold-blooded, coiled, and ready to strike at anyone within range.

"What will you do when you get out of the hospital?" I asked. I didn't say "if you get out." But Dwyer knew he was in trouble without my reminding him.

"I'll get light duty." Dwyer moved a hand, and the chains clicked like dominoes being shuffled. "Sweeping, handing out towels. Working in the commissary." He caught my eyes with his gaze and held them. "I'm going to join a Bible study group and attend chapel. I'm going to put my past behind me, give my life to God, and become a better person."

"Rob has been talking with Reverend Black since he was admitted," Joan said. "He stops by to pray with Rob several times a day."

"It keeps me from being afraid," Dwyer said.

Dwyer had the religious patter down, and maybe he was sincere. Yet I again suspected he was telling us exactly what he thought we wanted to hear. But didn't that, in a way, make him more human? He was a man pleading for his life.

"We're going to let you rest now," Joan said. She turned to me. "Unless there's something particular you wanted to ask."

I shook my head.

"Good luck." I considered shaking Dwyer's hand, but it would have been awkward. I was standing too far away. I gave him a quick wave and a small smile.

"Thanks." Dwyer lifted his hand to wave back. Once more, the chains clicked like dominoes when the links rubbed together.

The meeting on Dwyer didn't start until well past midnight. Five of us gathered in a large, shadowy room without a ceiling. Ducts and conduits ran directly overhead and, above them, skylights were visible as squares of blackness. Light fixtures hanging from cables produced circles of brightness that faded into gray at the fringes.

I shook hands with Tom Simon, a lean, almost gaunt, cardiologist in his late sixties. He introduced me to Rachel Hite, a middle-aged African-American woman who was dressed elegantly in a black suit with a flat white collar. She was an attorney from the hospital's legal department.

The fifth person was Bruce Mazer, a social worker in the transplantation center. He was in his thirties, short, with wide shoulders, and wearing a green tie with his black leather jacket. His dark hair was buzz cut, and two silver studs pierced his left ear. Both Bruce and Rachel were strangers to me.

"Let's get started." Joan was brusque, barely able to hide her impatience. She sat at the head of the table, with Rachel and Tom on her left. I was on her immediate right, with Bruce beside me. "Rachel, what's our legal status here?"

"The Center has no legal duty to accept Robert Dwyer as a transplant candidate." Rachel spoke crisply and looked fresh, as if accustomed to tense, late-night meetings. "A 1976 Supreme Court ruling held that states must provide prisoners with adequate medical care, and that's about all the law we have."

"Surely that means transplanting prisoners." Tom sat up straight and looked stern, as if delivering a judgment himself. "Mr. Dwyer has little heart function. While we can't guarantee him a heart, we can list him for one."

"California took that interpretation." Rachel shifted in her chair and smiled at Tom. "An inmate serving a fourteen-year sentence for robbery developed a heart-valve infection, and the correctional facility sent him to Stanford for evaluation."

"And they listed him?" Tom seemed surprised a center had actually done what he was proposing.

"He got a heart, and California DOC picked up the bill," Rachel said. "Two hundred thousand for surgery and hospitalization." She glanced at the yellow pad in a leather folder. "The estimated costs of long-term care was two million dollars."

The inflection in Rachel's voice suggested she was shocked by the figures. But the rest of us were familiar with the costs. People usually think only about the expense of performing a transplant, but total expenses accrue over years and include immunosuppressive drugs, lab tests, and the hospital stays required by episodes of acute rejection.

"That couldn't have been politically popular," Joan said. Light glinting off the oval lenses of her glasses made her eyes blank and unreadable. "And California is a liberal state."

"A bill was introduced to deny transplants to convicted felons," Rachel said. "But it didn't get anywhere."

"And I submit that it shouldn't have." Tom tapped the table to emphasize the point. "What about other states?"

"I bet the legislators were afraid they might end up in jail themselves, then need a transplant." Bruce spoke under his breath, addressing the remark only to me.

"Arkansas denied a liver transplant to James Earl Ray," Rachel said. Neither her tone nor expression revealed her attitude. "The state supreme court held that the denial did not count as cruel and unusual punishment. This was equivalent to ruling that no one has a constitutional right to a transplant."

"And Ray died of liver failure." Tom frowned. "I remember the case. Not many were sorry to see such a bad man die, but he might have lived another ten years with a transplant."

"And Rob Dwyer is the first case in this state?" Joan snapped out the question, calling us to focus on our task. "We haven't had one at our center."

"No, we have no precedent in the state." Rachel said. "Whatever you decide will set one." She smiled and gave a small laugh. "The DOC attorney made clear that they're happy to leave the disposition of the Dwyer case up to the center."

"Politics." Tom made the word sound vulgar. "No elected official ever wants to take the heat for a controversial decision."

"Nobody is going to complain if we don't transplant Dwyer," Bruce said. He cocked his head to the side and began rubbing his ear studs between his thumb and index finger. "The California dude was a robber. Dwyer is a killer, a double homicide."

"Wait a minute! Wait a minute!" Tom held up his hand. "The court sentenced Mr. Dwyer to life imprisonment, not to death from lack of medical care. And if he doesn't get a heart, he'll die."

"That's what he deserves." Bruce's voice rose in volume to match Tom's. "At least that's what people are going to say."

"We might be able to save this man's life." Tom scowled at Bruce. "That's the business we're in. So we need to give Mr. Dwyer a shot at staying alive."

"Not a particularly apt choice of phrase, Tom," Joan said. Her lips curled into a smile, and she seemed to be trying to keep from laughing. The rest of us, except for Tom, laughed out loud. Tom's face turned red, and he gave an embarrassed smile.

"What you want is so unfair, Tom." Bruce leaned across me, wanting to get closer to his adversary. "If John Q. Citizen needs a heart and doesn't have the cash or insurance to pay for a transplant, he doesn't get one."

"True," Tom admitted. "Not a heart."

Bruce jabbed his finger in the air. "Then along comes Dwyer, this baby killer, and not only is he going to get a heart, the state is going to pay for it?" Bruce shook his head violently. "Tell me I'm having a nightmare."

Tom leaned forward to respond, but seemed to think better of it. He limited himself to shaking his head, then lowering his eyes so he wouldn't have to look at Bruce.

"Let me summarize the discussion." Joan picked up the index card on which she had been taking notes. "Legally, Rachel says, we aren't required to put Dwyer on the transplant list." She glanced at Tom. "You think we ought to list him, because he could benefit from a new heart."

"It could save his life." Tom nodded at Joan, agreeing with her phrasing.

"And, Bruce," Joan said, "you think we shouldn't list Dwyer, because we'd be giving a killer medical care superior to the care poor people who aren't criminals could get." She looked at Bruce for confirmation.

"Yeah," Bruce said. "It burns me that the state's willing to save the life of a killer, but not the life of somebody who's honest but lacks money." He moved his chair away from the table with a sharp shove. "That sucks."

"Now I want to hear what Ron has to say." Joan took another index card out of the pocket of her white coat.

I'd been dreading the time when Joan would ask my opinion. I knew what I was going to recommend, but the decision didn't make me feel good. Could I make it clear that I was offering a reasoned judgment, not expressing a personal attitude?

"I don't like Rob Dwyer," I said. "He killed two people, and I find that reprehensible." My voice was tight, and sweat trickled from under my arms. "I also doubt he's changed much during his years in prison. He seems to think he's clever enough to trick people into letting him have his way."

"You nailed him." Bruce sounded eager to agree. "Like every sociopath I've ever met, he's totally manipulative."

"He's a killer and probably other bad things too," I said. "But he's still a person." I paused, thinking about how to make my point clearly. "And he's also a person in medical need."

"Yeah, right," Bruce said. "But his life isn't worth as much as the life of somebody who contributes to society." He sounded reasonable and neither smug nor spiteful.

"It is to him," Tom broke in. His voice had an edge of sarcasm. "And maybe to his parents and siblings."

"But our society is committed to the principle that everybody has an inherent and equal worth." I ignored Tom's comment. "So the life of a neurosurgeon isn't worth more than the life of a bus driver. Even if the neurosurgeon contributes something of greater value and would be harder to replace."

"But we're talking about a guy who *took* lives." Bruce dropped his tone of reasonableness. He sounded exasperated, as if I were missing the obvious point.

"That's why we're punishing him," I said. "We're depriving him of his liberty."

"The life of Robert Dwyer is worth the same as the life of anybody else?" Rachel looked across the table at me with a faint smile. She seemed more intrigued than puzzled.

"As a person, yes," I said. "Just the way we consider everyone equal under the law."

"So Dwyer is as entitled to a transplant as anybody else." Tom smiled broadly, as if he had produced a surprise.

"Dwyer needs a transplant and could benefit from one," I said. "That makes him qualified to be listed for a heart. Whether he gets one depends on one becoming available before he dies. He doesn't deserve special consideration, only equal consideration."

"So you're saying medical need and potential to benefit are the only two requirements." Joan spoke rapidly, pressing us to complete our discussion.

"Except for the third requirement." Bruce jabbed his finger at me. "It's also who can pay. Remember John Q. Citizen? He can't pay, but the state pays for Dwyer. How can you think it's okay to be unfair?" Sounding triumphal, he tossed his pencil on the table.

"The situation is unfortunate, but not unfair," I said. Sweat was running down my ribs and soaking my shirt. "I wish our society provided medical care to everybody who needs it but can't afford it." Tom and Joan nodded, knowing what it was like to turn away patients without resources. "But denying Dwyer a heart for that reason would be like refusing to feed him because we haven't agreed to feed everybody who can't afford to buy food."

"Arkansas didn't recognize a constitutional duty to care for James Earl Ray." Rachel offered the statement as a challenge. She wasn't hostile, only professional.

"Maybe the court was right about the Constitution," I said. "But on ethical grounds, the decision was wrong. If the state imprisons people, it assumes responsibility for their welfare. When we locked up Rob Dwyer, we simultaneously accepted the job of taking care of him."

"So, bottom line, we should list him for a heart?" Joan asked.

"Yes," I said. "I don't have to like him to think we have a duty to try to save his life by the best available means."

"Thank you." Joan put down her pen beside her index cards. "I've got everybody's arguments, so give me a minute."

Bruce raised his hand and started to say something, then changed his mind. We all sat quietly, focused on Joan. She removed her glasses and put them on the table next to her pen and cards. She massaged her eyes with her finger tips, then covered her face with her hands.

I glanced around, not looking at anything in particular. Bending my head back, I gazed up at the squares of blackness between the roof rafter. If we turned off the lights, could we see the stars? It occurred to me, ironic in the circumstances, that I was looking at one of the two things that Kant said filled him with awe—"the starry skies above me and the moral law within me."

Joan stared out at us for a moment, looking younger and less magisterial without her glasses. Her dark eyes shone. It was possible to see how she might have appeared as a medical student—intense and intelligent, yet not quite sure of herself. She picked up her glasses and hooked the legs of the frames behind her ears. The lenses flashed as the light caught them.

"I'm going to call UNOS and list Rob Dwyer as Status One." Joan's tone was brisk. She then let her breath out slowly, as if she no longer had to hold it in. "Let's hope we can him get a new heart."

"We can give him a shot at one," Tom said. He caught himself too late and grimaced, but none of us said anything.

"I thank you all for meeting at this inconvenient time," Joan said. "And I thank you for expressing your views so clearly. This was not an easy case, yet I think we've all learned something from it." The words were formulaic but sincere, like so much that physicians are required say.

We got up from our chairs and wandered toward the door. We seemed reluctant to leave, even though we'd completed our task. Joan hurried down the hall, presumably to get the data she needed to list Dwyer with UNOS. Tom went with her.

Bruce and I fell into step as we walked toward the elevator.

"You were just blowing smoke, weren't you?" he asked.

"What do you mean?"

"Deep down, you agree with me." Bruce stated it as a blunt fact. "When they're not being PC, everybody does."

"You want to know how I feel about Dwyer?"

"Go ahead and tell me." Bruce was smirking. He was sure he had my bleeding-heart-liberal number.

"I don't care if Dwyer dies," I said. "Even a slow and painful death is okay with me. And while he's dying, maybe he could watch a video loop of Kim and her little girl, so he could never forget that he killed them."

"Hey-sus Christos." Bruce sounded shocked. He frowned and looked at me with narrowed eyes. "But you didn't say anything like that at the meeting."

"I'm telling you how I *feel*." I turned and gave him a smile. "At the meeting I said what I thought was *right*."

Rob Dwyer got a heart matching his size and blood type within twelve hours of Joan's listing him. The transplant took place without complications, and by the end of the second week, Dwyer was sent to the prison infirmary to complete his convalescence. Joan told me that in ten more years Dwyer would be eligible for parole. He might, of course, need another transplant before then. Or he might die.

The United States is the only developed country where the issue of a convicted killer's qualifying for a heart transplant, when many law-abiding citizens don't, arises as a genuine problem. Elsewhere, the question is as hypothetical as whether transplants should be allotted to humans instead of Martians.

Other industrialized countries are committed to providing all their citizens with basic medical care. Thus, when it comes to qualifying for a heart transplant, whether in London, Paris, Rome, Berlin, or Toronto, the flower seller on the street corner is on equal footing with a neurosurgeon or a murderer locked away in prison. Medical need, not financial means, is the essential qualifying criterion. Money will not buy a ticket to a transplant, nor will criminal status keep you from qualifying.

What seems unfair about the state's paying for Rob Dwyer's heart transplant is that it isn't also willing to pay for Sue Citizen's, whose only crime is not having enough money or insurance to pay for it herself. This looks like a case either of rewarding criminals or punishing the needy.

I suggest that seeing the state's paying for Rob's transplant as *rewarding* him is like viewing an image in reverse. When critics complained to Clinton Duffy, the legendary warden of San Quentin, that he coddled convicts, Duffy's response was that people were send to prison *as* punishment, not *for* punishment. Rob Dwyer's punishment is being deprived of his liberty, not his life.

When we lock up prisoners, we prevent them from making most of their own decisions, including ones about their own welfare. We take control of their lives in a paternalistic way. We thus acquire paternalistic responsibilities, and these include seeing to their well being and health. We are obliged to protect prisoners from abuse, make sure they are fed a proper diet, have the chance to exercise, keep clean, and receive needed health care. (It is an irony of the American prison system that so many prisoners have their basic needs better met in prison than outside.)

UNFAIR?

The conclusion to draw from the case of Rob Dwyer, I believe, isn't that it is unfair for us to give prisoners the medial care needed to save their lives. We have a duty to try to do that.

The right conclusion, I think, is that it is unfair for us not to provide needed medical care to people on the outside who lack the means to pay for it. The answer to the question "Why should a convicted killer qualify for a heart transplant, while a needy citizen guilty of no crime cannot?" shouldn't be that the killer ought not qualify, but that the needy citizen should. To be blunt, *both* should qualify.

We live in the richest society in the history of the world, and we have no excuse for not adopting a health-care plan to guarantee all citizens life-saving treatments. Dozens of plans have been proposed during the last decade, and the great majority of the plans are insurance-based schemes, not "socialized medicine." We should see to it that everyone in our society has access to basic medical care, in the way everyone now receives basic education and basic police protection.

Does this mean that Sue Citizen would be guaranteed a heart transplant? No, because getting a transplant depends on the availability of a donor heart. But it does means her lack of financial resources won't rule her out as a candidate.

Sue Citizen would then be on a par with Rob Dwyer. The question of which one, if either, gets a heart transplant will be decided on the basis of such factors as their physical condition and the availability of a compatible heart. Thus, the only criteria to play a role will be ones that are medically relevant.

Any unfairness about the two cases can then be attributed to the workings of the universe. This is the same unfairness that manifests itself when wonderful people are struck down by deadly diseases, are killed in car crashes, or die from the lack of a donor heart. It's not the unfairness that comes from some having the money to pay for life-saving surgery, while others die because they lack the money. The first sort of unfairness is built into the nature of things and must be accepted (more of Virgil's *lacrimae rerum*). The second is the result of the way we run our society, and it can be altered any time we want.

EIGHT

The Last Thing You Can Do for Him

I MET GAIL LANDSTONE by arrangement at the botanical garden on a Wednesday afternoon in late April.

"Dr. Munson?" She was standing outside the tall wrought iron gates and took a step toward me.

"Are you Ms. Landstone?" I asked.

"Call me Gail." Her voice was flat and dead, and she didn't smile or extend her hand. "Ms. Landstone is what they call me at the hospital. I don't care if I ever hear it again."

"I hope I didn't keep you waiting." I was a few minutes early, but Gail seemed irritated.

"Not too long." She switched a small embroidered bag from her right hand to her left. "I had to get away, even if it meant leaving Davy. I promised Margaret I'd meet you."

She talked on without looking at me. "I don't see what good you can do, though. Maybe you can tell the doctors that Davy's not dead. That's what they said this morning. But they can't be sure of that, can they? Not really."

Gail was in her early thirties, an inch or two shorter than average, with a thick, stocky build. Her face was plump but drawn, and the skin around her eyes was puffy. Her dark hair was tangled, with small wisps hanging over her forehead. She was wearing tight black trousers and a loose rust-colored pullover. She wasn't bothering about her appearance, and I could understand why. If I were in her place, I might not be functioning at all.

"Let's find somewhere to sit." I gestured toward the open gate and the profusion of vegetation on the other side. "Someplace where we can talk privately."

Gail said nothing, but she walked with me along the brick path winding through the garden. She kept her head down, as if needing to focus on where to put her feet. She ignored the early blooming flowers and shrubs we walked past. They could have been ash heaps or piles of rusted metal.

The weather was cool and cloudless, and the clear light of the sun gave the colors around us an unusual intensity. The trees showed a blush of pale green as the tips of new leaves poked out of the buds on the bare, spindly branches. The sky was a brilliant watery blue, and the cropped grass on each side of the path had the dark green hue of artificial turf. Nothing natural looked quite real, and the garden seemed a world of fantasy and illusion.

"Gail says she can't think clearly in the hospital," Margaret Kleene had said when she called me that morning. "And you're both close to the botanical garden."

Margaret had been a student in my bioethics class in her nursing-school days. She had gone on to get a doctorate in nursing, something novel at the time, and she now devoted most of her efforts to clinical research.

"She needs to talk to somebody about her little boy, somebody who's not a physician," Margaret said. "She hates the doctors, and refuses to meet with the psychologist or the social worker. And she won't listen to me or any of the nurses, because she says we're too close to the doctors."

"What makes you think she'll talk to me?" I let my impatience with Margaret show. Gail was a stranger for whom I had no particular responsibility, and I had no wish to add another obligation to my already crowded day. "I'm a total stranger."

"But you'll understand her problem better than anyone," Margaret said. "She knows I trust you, so she'll give you the benefit of the doubt."

"What's her little boy's medical situation?"

"Oh, Dr. Munson, it's horrible." Margaret's whole manner changed, and the veneer of professionalism dropped away. "Head trauma. Severe head trauma. And he's only five." She choked on the last words, then stopped talking as abruptly as if she had hung up the phone.

Gail Landstone, I later learned, was Margaret's cousin, so Margaret had a family connection with Davy. Perhaps more important, she had a personal one. Neither Margaret nor Gail was married, and they lived in the same apartment complex. They spent a lot of their time off together, and they frequently spent it entertaining Davy. They went out for ice cream, took drives in the country, and picnicked by the swimming pool at the recreation area of their complex.

"Sorry." Margaret cleared her throat. "Davy was supposed to stay in the fenced play area while Gail went up to her apartment to start dinner. But somehow he got out the gate, then wandered along the alley to a house a few doors down." She paused, as if ordering the events in her mind. "A neighbor watched him climb on the railing around the basement stairwell, then she saw him fall. It's about ten feet down, the police said."

"Oh, my God," I said involuntarily.

"It's worse," Margaret said in a choked voice. "His head hit an iron shoe scraper set into the concrete." She started to cry, but she kept talking. "The scraper blade fractured his skull and drove bone fragments into his cerebral hemispheres. And of course it cut deep into his brain. The neighbor called the EMS, and when Davy arrived at the ER, the neurosurgeons got to work on him immediately."

"Did they give him a GCS score?"

"Three," Margaret said. The word was almost a sob.

Three was the lowest score possible on the Glasgow Coma Scale. It meant Davy had displayed no responses to even a sharp pain like being stuck with a needle. His eyes hadn't opened, and he hadn't moved his body. Someone with a low GCS score isn't likely to survive, no matter what kind of treatment he gets.

"When did all this happen?"

"Monday, Gail's day off." Margaret had pulled herself together and was speaking with professional briskness. "His breathing was irregular and his pupils fixed. It was clear he had a severe brain stem injury."

"Did he wake up after the surgery?"

"He's never been awake since he fell." Margaret took a breath, then sighed. "But, you know, kids are indestructible, so the nurses were telling Gail on Monday that maybe he might recover some functions. At least he might not die."

"Is anybody saying that now?"

"Not at all." Margaret sounded resigned. "Not with both hemispheres shredded and his brain stem damaged. The surgeons cleaned out the bone fragments, but they couldn't stop all the bleeding or get the swelling under control. Dr. Raymond's neurological team assessed him this morning and pronounced him."

Pronounced him, as in *pronounced him dead.*

"I'm sorry," I said. "You both have my sympathy."

"He's still on life support," Margaret said.

"Because he's a potential donor?"

"He can't be," Margaret said. "His organs were too damaged as a result of his injuries and their treatment."

"So has Gail asked that treatment be discontinued?" I was beginning to get an idea of why Margaret had called me.

"She's not rational," Margaret said. "Having your child die is horrible. It's terrible for me, and I'm not his mother." Her voice broke. "But Gail doesn't want to admit that he's dead."

When Margaret was my student, we had spent two seminar periods on defining death, and she knew that the issues are as much conceptual as medical.

"So what do you want me to do?"

"Talk to her," Margaret said. "Explain Davy's condition in a way that she can understand."

"You said she's not rational."

"Oh, I think that once she's had a chance to think things through, she'll change her mind"

"Don't be too sure," I said. "People tend to believe what they want to, especially when an issue is emotionally charged."

But I agreed to meet Gail and talk to her.

I guided her to a wooden park bench in a niche formed by a semicircle of forsythia bushes. She dropped heavily onto the seat and gazed toward the lily pond in front of us. A feathery plume of spray spouted from a fountain in the middle of the pond, and several white ducks glided across the smooth green water.

"I'm sorry about Davy," I said. "Margaret told me what happened to him."

"The doctors say he's dead." Gail spoke in a low, angry voice. "He doesn't look dead to me."

"I understand that his doctors performed all the standard tests, then concluded that his brain isn't working anymore." Directness seemed best. "They say his fall damaged his brain so badly it will never work again."

"But how can they know that?" Gail turned and gave me a hard, cold stare. "He might wake up in a week or a month. Maybe six months or two years. You hear about cases like that."

"Sometimes you do," I agreed. "But those are people who were in a coma. Their brains were injured, but not fatally. A few get better eventually and come out of the coma."

"But Davy is in a coma." Gail spoke triumphantly, as if scoring a point against me. "That's the first thing they told me when I arrived at the hospital on Monday afternoon."

"He was in a coma when he was first assessed," I said, not wanting to antagonize her. "Then when they did an MRI, they discovered his brain had been damaged extensively, and the surgeons confirmed this when they operated to remove the bone fragments."

"They said they didn't think surgery would help." Gail's face clouded, as if she were recalling the scene. "I insisted."

"I would have too." I nodded, understanding her desperation. "I'm sure they told you that the EEG they did this morning showed Davy's brain no longer has electrical activity. It's stopped functioning, and there's no chance he'll ever wake up."

I fell silent, wondering if I had been too blunt. Gail said nothing and kept her face turned away from me. She was scowling. I watched the ducks snapping at the barely visible insects skimming across the water. Their head movements were exceedingly swift. They were also surprisingly graceful. Yet who associated graceful movements with ducks?

"He's hooked up to machines." Gail gave me a sideways glance, as if daring me to disagree. She again switched her embroidered bag, moving it from her left hand back to her right. "They're keeping him alive."

"They're keeping his body functioning." I rejected her description. One of the ducks suddenly flapped its wings, gave a loud honk, and lunged at the duck beside it. The threatened duck swam away without a fight, then circled back to its original place. "That why Davy looks like he's alive."

Gail made no response. We sat without talking.

The world around us was beautiful. Painfully, ironically beautiful. The ducks, animated and dazzlingly white, swimming in the gray-green pond; the dark green of the floating lily pads; the sparkle of sunlight on the silvery spray of the fountain. The flowers on the long, drooping branches of the forsythia shone like burnished gold, and the sky was still that pale washed-out blue, with no sign of a cloud. Nothing should be wrong for anyone, but for Gail everything was.

"What is death?" Gail sounded angry but impersonal, as if addressing the universe, rather than me. "I've always wondered what happens after you die. Do you really have a soul that can go to heaven? But I've never asked when are you dead."

"It's a confusing question." I looked at Gail and saw that she was watching the ducks, perhaps so she wouldn't have to meet my eyes. "I'm not an expert on the topic, but I've given some thought to it."

The ducks were foraging for food. Some of them dipped their heads under the water to get to the lily-pad roots. They kicked their feet in the air. At another time with another person, I would have laughed at their antics.

"I can't handle a lot of complicated philosophy talk." Gail was curt, wanting to cut me off before I got started. "I can't follow it, and I'm not in the mood for it."

"Do you have a computer?"

"I've got one at work." She gave a sharp, humorless laugh. "You know what I do? I'm a debt collector. I'm always checking on people's accounts and payments, so I can call them up and put the squeeze on them."

"Maybe this comparison will help," I said. "If you erase all the programs installed on your computer, the only thing remaining on the hard drive is the operating system."

"Okay." She spoke with hesitation, as if expecting a trick.

"The program files are gone, so you can no longer access their contents or add to them."

"Yeah, I got it." She sounded bored.

"Some claim that's what happens when the hemispheres of the brain are severely damaged. The injured person has no self-awareness, can't take in any information, and isn't able to respond to anybody." I paused to give her a chance to think through the analogy. "So according to this view, death occurs when people lose these higher brain functions. Without those functions, they stop being people."

"Like that poor woman who lived so long." Gail still spoke abstractly, but the anger had gone. "The one whose parents had to go to court to get her out of intensive care."

"Karen Quinlan," I said. "Probably that's who you mean, but another case talked about a great deal was Nancy Cruzan in Missouri. A Supreme Court ruling permitted her parents to discontinue her life support."

"I don't remember her." Gail sounded curious.

"The cerebral hemispheres of her brain were damaged in a car wreck," I said. "She was left in what doctors call a vegetative state. She wasn't conscious or aware of anything."

"And is that what they say Davy is in?" Gail turned to look at me. Her eyes were dry, but her face was crumpled into a deep frown, as if she were in utter misery.

"No." I shook my head. "Nancy Cruzan was like Karen Quinlan. Both had an undamaged brain stem that kept their bodies operating. Their hearts pumped blood, and they could breathe on their own, even though they lacked the capacity for thinking or responding to people or the world around them."

"So their operating systems were functioning, but their brain programs were wiped out."

I was briefly confused, then I realized Gail had taken my analogy to the next step. She was ahead of my explanation.

"That's right," I said. "But Davy no longer has an operating system. You might say he doesn't even have a hard drive, because his brain stem was terribly damaged. His blood flows only because a machine pumps it, and his lungs inflate only because a machine forces air into them."

Gail gazed out at the pond. The ducks were still diving and snapping at bugs, but she showed no sign of watching them. Perhaps their existence didn't register on her consciousness.

"His skin is so warm and soft." She spoke in a musing tone. "Except for the tubes and wires, he looks the way he does when he's asleep." She paused. "It's true he doesn't move. Usually when he sleeps, his hand twitches, or his legs. Sometimes he kicks his feet, like dogs do when they're dreaming."

"I'm very sorry," I said gently.

"It's hard to believe." Gail made a sort of choking sound. "It's so hard I can't do it. I can't make myself."

"Maybe it's easier when machines aren't involved," I said. "Easier on the survivors. People are just as sad and upset, but the course of nature seems clearer."

"What happens when they turn off the machines?"

"You'll see that Davy is already gone," I said.

"I guess that's what I'm afraid of." Gail spoke in a faint, whispery voice.

"It's like seeing that the magician palmed the coin." I ignored her comment. "The vanishing was only an illusion."

"You're saying that seeing Davy looking like he's alive is only an illusion." It wasn't a question. "It's a trick of the machines."

"Along with a reluctance to admit that the worst has already happened," I said. "That even hope is pointless."

"Hope is pointless," Gail repeated.

We sat together, each with our thoughts. On the far side of the pond, a little girl in a white dress and pink sweater began throwing pieces of bread into the water. The ducks produced a sudden din of quacking and swam toward her. Even ducks farther down the pond fluttered their feathers and began swimming for the floating bread.

"I need to go back to the hospital." Gail turned toward me. "Will you go with me?"

I was on the verge of inventing many excuses. But I had gone too far with Gail to turn away. It would be too much like abandoning her.

I had forgotten that five-year-olds were so small.

David Landstone was hardly three feet tall. He looked tiny in the hospital bed. A child-sized endotracheal tube was in his throat. Another, smaller, tube ran down his nose and into his stomach. Two IV lines dripped solutions into his arms, and the leads from monitors ran across the sheet. The main sound in the room was the steady *whoosh* of the ventilator.

Davy's cheeks were chubby, and his nose was tilted and sharp. Otherwise, it was difficult to see what he looked like. His eyelids were taped shut, and his head, from the eyebrows up, was swathed in white bandages. Near the lobes of his ears, a few strands of hair protruded from under the bottom layer of gauze. His hair was as inky black as his mother's.

I stood with Gail beside the bed. Dr. Alan Altzer and Mary Jacobson, a critical-care nurse, were crowded into the side of the room between he wall and the bed. Dr. Altzer, a short, gnomish man with a close-cropped black beard, was reading through Davy's chart. Mary Jacobson was entering numbers into the bedside computer terminal.

Gail put her hand on Davy's arm. She stroked his skin as if she were soothing him, calming him down from some upset. She kept her eyes fixed on his face.

"If you wait outside a few minutes, we'll get Davy ready," Dr. Altzer said to Gail. She hadn't told him she wanted the lines and tubes removed, but he seemed to realize she had come to a decision.

Gail looked at Dr. Altzer for a long moment, then nodded. I followed her into the small carpeted waiting area. Two people were there, a young couple. The man was in work-clothes, green twill pants and a shirt with "Bob" on a white patch. He looked up at us with a tense smile, but the woman kept her head lowered. It was impossible not to wonder what sort of crisis they were facing with their child.

I nodded to Bob, but Gail and I didn't sit down. We stood beside the half-wall separating the waiting area from the hallway leading to the rooms.

"He looks peaceful." Gail spoke softly, not addressing me, but apparently reassuring herself.

"He does." I matched her soft tone, not wanting to say or do anything that would upset her more.

Suddenly, without saying anything, Gail turned and walked quickly down the hall and back into Davy's room. I assumed she had changed her mind and wanted to be at Davy's side when the equipment was turned off, and I didn't want to violate her privacy. Then I heard Gail's raised voice. I couldn't make out the words, but she was agitated and angry.

I decided to see if I could help.

Gail and Dr. Altzer, both short, were standing face-to-face. Gail's round, pudgy face was beet-red, and her hair seemed even more disheveled. Dr. Altzer's face was frozen into impassivity, but his eyes looked fierce. The nurse stood beside the respirator at the head of Davy's bed. She was watching like a stunned spectator at the scene of an accident.

"I'll sue the hospital, and I'll sue you personally." Gail's voice was harsh, and she jabbed a finger at Dr. Altzer.

"I'll call a lawyer and get a court order."

"You'll have to do what you think best." Dr. Altzer spoke with deliberate control. "It's hospital policy to stop treatment when a patient is declared dead. I thought you understood that and agreed with it."

"I changed my mind." Gail turned and pointed at me. "He confused me with talk about computers and programs. For me, Davy is alive, and I'd rather have him alive this way for years than have him dead forever."

"But he's not alive." Dr. Altzer sounded more understanding. "I explained brain death to you, and it sounds like Dr. Munson did too."

"I think it's time to let Davy go." I said it gently. "You know he wouldn't want to stay hooked up to all this equipment for months or even years."

"I don't know what to do." Gail's covered her face with her hands. Her words were muffled, and her shoulders shook with sobs.

"Let him go." I said it with more force this time. "A boy who climbed a gate to go exploring shouldn't have to linger like this. This is the last thing you can do for him."

Gail drew in her breath in sharp hisses, trying to control her crying. If she had heard me, she gave no sign. I decided I had said enough.

Over Gail's shoulder, I saw four uniformed security officers standing outside the door. Dr. Altzer or Mary Jacobson must have called security when Gail started making trouble. Dr. Altzer also noticed the officers. He held up the flat of his hand to order them to keep out of the room.

"Just stay here a moment, and we'll get Davy ready for you to say goodbye." Dr. Altzer spoke briskly, as if by being businesslike he could keep Gail from thinking too much about what he was going to do.

Dr. Altzer nodded to Mary. She immediately held down the safety switch on the ventilator, then turned it off. She peeled off the tape holding the endotracheal tube in place and tugged the tube out of the boy's throat. She wrapped a towel around it so that the secretions wouldn't drip on him.

While Dr. Altzer pulled out the IV lines, Mary Jacobson wiped Davy's face with a wet washcloth, then used alcohol pads to clean the sticky reside of adhesive tape off his skin. She and Dr. Altzer straightened out Davy's body and tucked the sheet around him, leaving his hands on the outside. Mary lifted the little boy's head and fluffed up the pillow.

Dr. Altzer walked over to Gail and put his arm around her shoulders. She stood with her hands over her face, not wanting to see anything. She was no longer sobbing, but tears had leaked through her fingers, wetting her hands.

"He's ready." Dr. Altzer spoke with urgency, as if someone were waiting for her. Gail lowered her hands, then blinked in the light. Dr. Altzer said, "Wipe your eyes and blow your nose."

Gail took the wad of tissues he held out. She did as he told her, making a loud honking noise when she blew her nose. I thought of the squawking of the ducks in the botanical garden.

Dr. Altzer guided Gail to a chair the nurse had placed at the head of the bed. Gail bent over and kissed her son's cheek. She then sat down and picked up his hand. She gently stroked it, but she kept her eyes on his face.

Davy's skin had looked pink and healthy, but now it was visibly changing. The rosy glow was fading to the pale gray, almost white color of cigarette ashes. His blood was no longer circulating, and like a receding tide, it was draining out of his tissues. Under the pull of gravity, the blood would eventually settle in the parts of his body nearest the ground. This was the lividity, the dark purplish color, medical examiners look for to help them determine the time of death.

Lividity would be misleading in Davy's case. He had died that morning just after Dr. Raymond reviewed his medical history, examined him, and evaluated the EEG data. Davy had died, legally and conceptually, when Dr. Raymond exercised his clinical judgment and pronounced him dead.

Gail was still stroking Davy's hand when I left the room. I knew she would never want to see me again. I was a stranger, and I had witnessed the worst thing that would ever happen to her.

Margaret Kleene called me late that evening to thank me for meeting with her cousin.

"I knew once she talked to you, you'd change her mind about brain death," Margaret said. "She accompanied Davy's body to the funeral home, so now the healing process can start. And not just for her, for me too."

"I didn't change her mind," I said. "I told you that people tend to believe what they want to when they're in an emotional state. And there's nothing more emotional than having your child die."

"Gail didn't believe what you told her?"

"Probably not," I said.

"But she let them take Davy off life support." Margaret sounded puzzled.

"I made an appeal to her emotions, instead of to her reason." I thought back to the scene beside the bed. "I told her that letting Davy go was the last thing she could do for him."

"What's wrong with saying that?"

"Because Davy was already dead," I said. "She couldn't do anything to help. But I thought if she believed she could, she would find it easier to accept his death."

"I told you Gail wasn't in her right mind." Margaret's tone was earnest. "You couldn't have persuaded her."

"Perhaps not," I admitted.

"And I know hospital policy as well as you do," Margaret said. "They were going to take Davy off life support today, whether Gail agreed or not."

"That was the one favor I could do for her," I said.

"What was that?"

"I kept her from being thrown out of the hospital by the security people," I said.

What I didn't tell Margaret was that to achieve this I had to let Gail believe I accepted her view that her child was still alive. I had been, to that extent, dishonest and manipulative. Yet, considering the alternative, I forgave myself.

Unlike our computers, our bodies aren't equipped with a tiny light that stops glowing when they stop functioning. Thus, no society has ever been quite sure when to declare someone dead. Some cultures in the Middle East don't consider people dead until three days after their hearts stop beating.

Gail Landstone was faced with a situation in which, in outward appearance, her child was alive. Davy's blood was circulating, his temperature was normal, and his chest moved in and out. Yet what she saw was an illusion of life created by drugs and machines. Fifty years earlier, Gail would never have had to deal with the emotional and conceptual challenge of accepting that Davy was dead and that stopping his treatment was the right thing to do.

DEATH AS A PRACTICAL MATTER

"When is someone dead?" didn't become a practical question until the rise of intensive-care medicine in the 1950s and increasing success of organ transplantation in the 1970s. People began to ask, "If a physician switches off the ventilator keeping a patient's body supplied with oxygen, is that homicide? If a surgeon removes the heart from a breathing patient, has she killed him?"

These questions became more than academic exercises for a few surgeons who were arrested and charged with homicide, and finding answers became more urgent for personal and practical reasons. If surgeons couldn't remove organs from a heart-beating body without fearing a trial and a prison sentence, they would no longer perform transplants.

Discussions during the 1970s and 1980s about determining death led to the development of four basic ideas about what it means to be dead. These remain the definitions at the focus of current debates.

1. Cardiopulmonary

A person is dead when her heart stops beating and she is no longer breathing. This is the traditional definition of death as "the permanent cessation of breathing and blood circulation." It is akin to what people have in mind when they say, "He died twice while they were operating on him."

2. Whole Brain

Death is "the irreversible cessation of all brain functions." A person is dead when his brain displays no organized electrical activity, and even the brain stem, which controls basic functions like breathing and blood pressure, is electrically inactive.

3. Higher-Brain

Death is the permanent loss of consciousness. An individual in an irreversible coma is dead, even though her brain stem continues to regulate her heartbeat and blood pressure.

4. Personhood

Death occurs when someone ceases to be a person. This definition is concerned with the absence of mental activities like reasoning, remembering, experiencing emotions, anticipating the future, and interacting with others.

DEATH AS A DIAGNOSIS

The definition of death in the 1985 federal Universal Determination of Death Act is a straightforward endorsement of concepts one and two:

> An individual who has sustained either (1) irreversible cessation of circulatory and respiratory functions or (2) irreversible cessation of all functions of the entire brain, including the brain stem, is dead.

This definition is embodied in the laws of all fifty states, but that we accept both concepts can be confusing. Like Gail Landstone, many people don't see how someone can be declared dead whose body is still working. But many others don't see how someone can be declared dead unless a doctor tests him to make sure he is brain-dead.

The key to eliminating such confusions is recognizing that death is a diagnosis governed by two sets of criteria. In some circumstances, cardiopulmonary criteria are appropriate, while in others, brain-death criteria are. Neither set trumps the other.

DIAGNOSIS

A physician makes a diagnosis by confirming the hypothesis that the patient's disorder best fits into a particular category. Thus, a five-year-old girl who has a fever, sensitivity to light, and a red, pustular rash has chicken pox. Symptoms and signs define the "chicken pox" category, and the data about the little girl confirm the hypothesis that these criteria are met.

So it is when "death" is the diagnostic category.

CARDIOPULMONARY

The cardiopulmonary criteria dictate that a patient is dead when his circulatory and respiratory functions have irreversibly ceased. To determine if the data support this hypothesis, the physician examines the patient. She may feel for a pulse in the carotid or femoral artery and use a stethoscope to listen for heart and lung sounds. She may use an ophthalmoscope to see if the blood in the vessels of the retinas has broken into the stagnant segments, indicating that the blood isn't circulating. (This is called the *boxcars sign*.) She may also perform an electrocardiogram to determine if the heart is displaying any electrical activity. Using such data, the physician may conclude that the cardiopulmonary criteria are satisfied.

Death is her diagnostic conclusion.

WHOLE BRAIN

The diagnosis of death using whole-brain criteria follows the same logic, but two restrictions govern their use. First, the criteria don't apply to anencephalic infants or children under two. Anencephalic infants are born without brain hemispheres, and because they lack even the potential for consciousness, it makes no sense to test for its loss. Also, young children develop at such different rates neurologically that clinical and imaging tests can't be used to make reliable predictions. For these groups, only cardiopulmonary criteria are appropriate for determining death.

Second, before a physician pronounces someone in a coma dead, he must rule out reversible causes. He must establish that the coma was not caused by drug use, an internal chemical imbalance (as in diabetes), or hypothermia. Patients with these conditions may show clinical signs of death, yet recover consciousness. For example, *fugu*, or puffer-fish liver, is a delicacy in Japan, but if a diner eats too much of it, its poison (a tetrodotoxin) may induce a state in which the diner's pulse and respiration become too slow to detect. Some victims of *fugu* poisoning have awakened to find themselves naked, cold, and shivering in a morgue. (It seems likely that other *fugu* victims didn't wake up at the right time and suffered an accidental death at the hands of mortuary workers.) Thus, it's crucial for the examining physician to be sure that a patient hasn't consumed something that might cause a reversible coma.

Once these preliminary conditions are met, the physician examines the patient to determine if her brain stem has suffered irreversible damage. She is removed briefly from a ventilator, at least once and often twice, to see if an increase of carbon dioxide in her blood will trigger her body's attempt to breathe. If it doesn't, this indicates damage to the brain stem.

The physician will also test other reflexes controlled by the brain stem. He strokes the back of her throat to check her gag reflex, shines a light in her eyes to see if her pupils contract, touches her eyeball to test for a blink response, and sticks her with a needle to look for a pain response. He may turn her head to the side to see if her eyes move with it—the *doll's eyes sign*.

Clinical observations like these are the primary data used to determine death. Often, however, a physician will order an MRI or a CT scan in an attempt to look for the amount and location of the brain damage, and an electroencephalogram can be used to confirm the clinical judgment that the patient's brain has undergone irreversible structural damage. (Even an irreversibly damaged brain will probably show some electrical activity, because isolated groups of cells remain active, but there must be no pattern of organized activity.) These tests are most likely to be used when a patient is young or was not expected to die from his injury or disease.

The physician relies on these data to decide whether they support the hypothesis that the patient's brain has suffered an irreversible loss of all functions. If so, death is the diagnosis.

SUCCESS

The cardiopulmonary criteria in the Uniform Determination of Death Act are traditional, but the brain-death criteria are modeled on those in the 1968 report of the Harvard Medical School's Ad Hoc Committee to Examine the Definition of Brain Death.

The committee was explicit about why a definition was needed. "Our primary purpose," the report stated, "is to define irreversible coma as a new criterion for death," because the cardiopulmonary criteria "can lead to controversy in obtaining organs for transplantation." Also, the committee wanted to avoid wasting resources on patients unable to benefit from life-support measures and to spare their families the financial and emotional costs of supporting them.

More than 99 percent of the people who die in hospitals are pronounced dead by traditional cardiopulmonary criteria. Those declared dead by brain-death criteria are always patients receiving intensive care. Once they are declared dead, they are taken off life support or, with the consent of their families, their organs are removed for transplantation.

The Harvard committee was successful in both its aims. The concept of brain death freed many families from the doubt and guilt involved in deciding to remove someone from intensive care. The concept also improved the success rate of transplants by allowing surgeons to use undamaged organs from bodies kept functioning by intensive measures.

HIGHER BRAIN FUNCTIONS AND PVS

Some would like to see society adopt the third concept and define death as the "irreversible loss of higher brain function." This would expand the criteria for determining death to include those in a persistent vegetative state (PVS).

Those diagnosed with PVS have damaged cerebral hemispheres, and this results in their not being aware of themselves or their surroundings. They are incapable of thinking or intentional movement, but if their brain stems are undamaged, their autonomic nervous system continues to control their reflexes.

Thus PVS patients can breathe and excrete, their hearts beat, their muscles respond to stimuli, and they cycle through regular sleep-wake patterns. Although their eyelids may blink and their eyes move, they lack the brain capacity needed to see. They are like digital cameras with a functioning optical system but no microprocessor—information is supplied, but it can't be used. Some PVS patients may smile or produce tears that run down their cheeks, but these are reflexes that are only accidentally connected with what's happening around them.

After six months to a year, PVS patients are not likely to recover even the most rudimentary form of consciousness. They aren't like patients diagnosed as "minimally conscious," who have some episodes of awareness and a small, yet real, possibility of waking up. PVS patients remain vegetative for as long as they live, which may be decades. Karen Quinlan lived for almost ten years, and Nancy Cruzan was allowed to die after seven. Terri Schiavo slipped into a coma in 1990 and died in 2005, only after a protracted legal battle by her parents to keep her husband from ordering her removed from life support.

PVS patients require total care. They must be fed through a surgically implanted gastric tube, hydrated with IV fluids, bathed and toileted, kept on special mattresses and turned to avoid pressure sores, given antibiotics to prevent infections, and provided with around-the-clock nursing care.

If the "irreversible loss of higher brain function" were accepted as the third legal definition of death, PVS patients could be declared dead and no longer given life-support measures, including gastric feeding and IV hydration. This wouldn't require a court decision or even a request from the family.

Adopting this criterion would mean that as many as fifty thousand PVS patients in the U.S. could be removed from life support. (No one is sure of exact number of PVS patients.) This would result in immense savings, because it costs about one hundred thousand dollars a year to provide care for a PVS patient. This money, proponents of adopting the criterion argue, could be better spend on extending the lives of those who are conscious and play a role in the lives of their family and society.

UNLIKELY

The whole-brain definition of death was accepted in our society with little fuss. To most people, it made intuitive sense that someone whose brain isn't functioning at all and will never function again can't be alive. The criterion also seemed cut and dried: the EEG shows that the brain is electrically silent.

The higher-brain definition is unlikely ever to be accepted. It lacks the precision and certainty that make the whole-brain definition uncontroversial. Given the same data from brain scans and clinical tests, doctors can disagree about when higher-brain functions have been lost permanently. Studies show, further, that PVS is frequently misdiagnosed, so many patients capable of recovery might fall victim to a bad diagnosis.

Also, the families of some PVS patients believe that, no matter how long the odds, the patient may eventually wake up from the coma. The thinking of these families is also probably representative of a large part of the population, even those who know no one with PVS.

LOSS OF PERSONHOOD

The definition of death as the loss of personhood has a resonance with most people. We understand what a friend means when she says her mother's progressive dementia has so destroyed her as a person that she might as well be dead.

We see how a woman with a degenerative neurological disorder like Alzheimer's or Huntington's may reach a later stage where she has lost so much cognitive and emotional functioning that she may no longer be thought of as a person. If we could agree that she has lost whatever is essential to being a person, given the "loss of personhood" definition of death, it would then be morally permissible for us to withdraw life support.

Understandable or not, the definition isn't likely to gain the support it would need to be adopted as a legal criterion. Exactly what attributes are required to qualify as a person is open to dispute, and we're unlikely to get general agreement on exactly when someone has lost so many of them that she has lost her status. Diagnosing death might thus come to be seen as arbitrary. Worse, it could become arbitrary, opening the road to abuse. The old, the poor, and the poorly functioning might stop measuring up to personhood and not be given needed support.

A large number of people will always believe that an individual they care about, no matter how mentally and physically impaired, is the same person as before. Looking like a person and having a history as a person are seen by many as sufficient for being a person. This is not likely to change.

CIRCLE COMPLETED

We end where we began: death can be defined as the permanent failure of heartbeat and respiration or as the permanent failure of the whole brain. These definitions have served us well. That they are seen as objective and precise gives them a strength that the other two definitions can't match. Leaving well enough alone seems to be our best option.

We can hope that as the concept of brain death becomes more familiar, people like Gail Landstone won't have to suffer the pain of an unrealistic hope based on doubt. Whether diagnosed by cardio-pulmonary criteria or whole-brain criteria, the diagnosis is the same.

NINE

The Boy Who Was Addicted to Pain

CHASE GRANGER ARRIVED home from school around four-thirty. The moment he walked through the back door and dropped his school books on the kitchen table, he felt a small thrill of anticipation. He knew exactly what he was going to do and was eager to begin.

He had started thinking about it during his one o'clock class. He'd tried to pay attention to what Mr. Kalten was saying about German economic conditions causing World War II, but he'd felt sick to his stomach. His mind kept returning to what Lopez had said and done. Yet he couldn't bear dwelling on them, so he'd nudged his thoughts onto a familiar path.

He imagined what he would do when he got home, going step by step and lingering over the images as they took shape. He saw himself from the outside, as if he were looking through the lens of a movie camera. He shot a wide angle to frame the scene that showed him in the bathroom, then pulled in for a tight shot of his almost-naked body. He finally narrowed the focus to an extreme close-up of his right thigh.

Then, suddenly, as if a switch had been flipped, he became the person he'd been watching. He felt the coolness of the pale green tile, the sweat of anticipation, the pain of the blade. The blood oozed, then trickled. Pain and relief existed in the world his mind had called up, but Lopez and the College School didn't.

His upset stomach soon started to feel better.

The morning at CS had passed as usual. The Populars in the hall had given him tight, polite smiles, and a few had said "Hi, how's it going?" They had been in classes with him, so they couldn't help recognizing him. Yet some, mostly girls, pretended not to see him and walked past as if he were invisible. They made him feel like the dead kid in *The Sixth Sense*, the one who couldn't be seen by living people.

He wasn't the only one at CS the Populars couldn't see, but he was one of the few who didn't fit into any group. The Wannabes saw the Populars, and the Populars saw them often enough to keep up their hopes. The Clubbies overlapped both groups, so everybody saw them. Only about six people in his class of ninety were as socially dead as he was.

He'd never been sure exactly what was wrong with him, but he knew his clothes were wrong. He wore the usual jeans, shirts, and shoes. But his shirts and shoes were the wrong style, out of fashion, or the wrong color. His jeans were made by the wrong company or had the wrong cut.

He sometimes could hear Populars giggle when he walked past. Did they think he was so stupid that when a guy asked, "Where did you get that cool sweatshirt?" and it had *Gap* on it in huge white letters, he wouldn't understand they were putting him down?

No, they *knew* he'd understand. They wanted to send him the message that he was uncool without saying it. Saying it would be rude, and at CS being rude was uncool.

He had tried to get his clothes right since seventh grade, but now he realized that, even if he succeeded, it wouldn't make any difference. He was on a scholarship, and his parents earned little money, didn't belong to a country club, didn't travel to Paris, or fly to New York or L.A. on business. He was uncool in the way that a dog was a dog. It was a permanent condition, a status he couldn't change.

The break between lunch and class was when things had gone wrong. He had been in the brick plaza outside the dining hall with Sam and Henry discussing the Debate Club. Sam was a Popular, but he was president of the club, so it was okay for him to talk to the likes of Henry and Chase when conducting business. While the three of them were planning practice sessions, Marc Lopez strolled over. Marc was short and squat, with thick shoulders and black hair cut so short on the sides his scalp showed white. He was a thug, but the Populars sucked up to him.

Maybe it was because his family was outrageously rich. Or maybe the Populars were scared of him. People said Marc had been kicked out of La Jolla Country Day in tenth grade because he cut up a kid. They also said he still carried a flip knife. That was possible, because the school had no metal detectors and nobody was ever searched. CS counted on the upper-class backgrounds of its students to keep drugs and weapons out of school.

Marc had walked up and said, "Hey, Sam, what's with wasting time with losers?" He glanced at Henry, then stared hard at Chase, putting him under the microscope.

"Debate business," Sam said. He took a step back, easing away from Henry and Chase, yet not getting closer to Marc. "Big tournament coming up."

"You ought to win easy," Lopez said. He narrowed his eyes into a squint. "When they see the big beak on this fat fart, they'll get scared he'll peck off their nuts and give up." He gave a braying laugh. "Big Bird wins for CS!"

Sam glanced away to avoid meeting Chase's eyes. Henry said nothing, but he looked relieved that Lopez hadn't made him the target.

The joke was stupid. Yet Chase felt his cheeks grow hot. He lowered his eyes. He knew he was overweight, but he was most self-conscious about his nose. He'd tried telling himself nobody noticed it, but he'd never be able to do that again. Lopez made it clear that people saw him as a fat freak.

Then Lopez had spit on him.

Not *on* him exactly. Lopez spit at the ground, and a blob of thick slime landed on Chase's right shoe. You couldn't prove it was deliberate, yet they all knew it was. Sam and Henry pretended not to notice. Lopez grinned.

Chase's heart gave a heavy thump, and for a moment his vision blurred. He should punch Lopez in his stupid, grinning face. If he did, Lopez would hurt him and might stab him.

He hesitated.

In that instant, the moment passed before he realized it was going. Taking a deep breath, he blinked. He should have hit Lopez without pausing to think. That he hadn't proved how pathetic he was.

Then the bell rang, and they headed toward Langston Hall. Lopez walked beside Sam, and Henry rushed ahead. Chase didn't try to catch up. That would only add to his humiliation.

He stopped by the restroom and wiped the spit off his shoe with a wad of paper towels. He then scrubbed his hands with soap until they were red. In class, he tried to concentrate on Mr. Kalten's account of *Kristallnacht*, and that's when his mind kept returning to Lopez. He saw the blob of spit, heard Lopez's laugh, and saw Henry's sickly look of shock.

Sitting in class, he found himself picturing different versions of the events. In one, he flattened Lopez with a punch to the jaw, then resumed his conversation with Sam and Henry. In another, Lopez pulled a knife, and he whipped out a pistol from his back pocket and shot him. He liked watching Lopez's smug expression change into a look of disbelief and fear before he collapsed to the brick pavement.

Chase sat through algebra and chemistry in the same distracted daze. Yet once he'd slipped into his dream mode, he started to feel better. The images of the steps he'd take—the blade along the skin revealing a white line, the blood oozing, then trickling—were soothing. They shouldered aside the disturbing images—the insult, the blob of slime, the bleak sense of shame.

Chase had discovered a few months ago that playing the images in his head gave him almost as much pleasure as experiencing the real thing. Except the images tended to grow faint and had to be refreshed every few weeks.

Chase was the only one in the house. Because he was seventeen, his parents considered him mature enough to stay alone. Charlie, his younger brother, was twelve, and he had to remain in the after-care program at his public school until their mother picked him up on her way home.

She worked at a community food bank, filling the grocery orders of people qualifying for assistance. Sometimes she was able to bring home donated food like smoked oysters or falafel mix, things that people on assistance didn't want. His mother accepted anything she was given, because their family was only a step or two above qualifying for assistance themselves.

His dad was employed as a painter and drywaller by a company that fixed up repossessed houses for resale. He worked long hours, not getting home until seven or eight at night. He didn't complain, though, because jobs were scarce and the family needed his overtime pay.

After dumping his books, Chase hurried down the hall to the bedroom he shared with Charlie. He took off his shoes and socks, then removed his shirt and threw it on his bed. He pulled open the bottom drawer of the varnished pine bureau. The drawer was jammed full of jeans and sweatshirts, and he ran his hands under the clothes until he felt the shape of his X-Acto set.

The flat wooden box had tiny brass hinges. Inside, an aluminum blade holder with a crosshatched collar rested in a notched frame, and four thin steel blades occupied a double row of slots. The holder was the diameter of a ballpoint pen, and the tip of each blade had a different shape.

He selected the blade in the first slot, because its stiletto tip made it perfect for his work. He slid the blunt end into the holder, then tightened the collar. He put the box back in the drawer, hiding it under the clothes again.

He went into the small bathroom off the hall and locked the door. When he heard the deadbolt click, he felt safe for the first time that day. Sweat tricked from under his arms, but the pale green tile floor felt cool on his feet.

The tiled walls and dark green porcelain fixtures gave the tiny room an antiseptic look. *Like an operating room*, he thought. He got the bottle of alcohol out of the medicine cabinet and unscrewed the cap. The alcohol was also green, dyed that color so winos would know that if they drank it, they would go blind. Despite the dye, the alcohol had the same doctor's-office odor.

Holding the X-Acto knife by its rounded end, he lowered the blade into the alcohol and swished it around. He placed the knife on the wash basin so the blade stuck over the edge. He thought of the procedure as "using sterile techniques."

He pulled off his jeans and kicked them into a corner, but he kept on his white knit briefs. He put down the toilet seat and sat on it. He raised his right leg and placed the side of his right foot on his left thigh. This exposed the smooth expanse of skin on the inside of his right thigh. A large portion of the area was as flat and tight as a stretched canvas.

He felt a thrill of excitement as he picked up the knife. He held it like a pencil, his fingers gripping the collar, then he started to work. He made a quick jab with the stiletto point and felt a sudden give as the skin split. The sharp pinch of pain made him suck in his breath.

He drew the blade along his thigh, cutting a two-inch line. The edges of the skin were white, then bright red blood welled up, filling the cut and blurring the line. He placed the blade tip on the original incision point, then sliced another two-inch line. The two lines intersected at a thirty-five-degree angle, and his third cut completed the triangle.

He paused to consider his work. The X-Acto blade was so sharp that it blazed through his skin with the precision of a laser. As he watched the blood ooze from the incisions, he felt himself relax. The tension went out of him as if a taut wire buried deep inside had been severed. He smiled and gave a little laugh. He felt pleased with himself.

His work wasn't finished. He pulled out a strip of toilet paper and folded it into a pad. He blotted the bleeding lines, then wiped the skin around them to keep the blood off the floor. He wouldn't know what to say if his mother asked him where the blood in the bathroom came from. He would have to lie, but he didn't like lying to his parents.

After repositioning his heel to stabilize his thigh, he cut the second triangle. He oriented it so its apex pointed inward, and the two triangles formed parts of a pentagram. He used more toilet paper to soak up the blood from the new lines.

The cuts were painful. He didn't find pleasure in the pain, but he found the cuts satisfying. Not the cuts themselves so much as the cutting. Cutting himself didn't make him happy, but it raised his spirits. He felt alive and excited, and the deadness he'd been enduring had disappeared like a cured headache.

The cuts were also beautiful. He liked the way the blood sprang from the smooth surface of his skin, then seeped and trickled across his thigh until his body finally stopped the bleeding. That he made the cuts himself was also something he liked. He caused the pain, and he experienced the feeling. He was in control.

Because the lines didn't go away for a few weeks, they served as a reminder. He could look at them and remember the stab of pain when he inserted the point, then the steady burn when he drew the blade along the blank paper of his skin. For two or three weeks, a glance at the cuts could raise his spirits.

He felt enough better that he was looking forward to doing his history reading. Unlike Mr. Kalten, he suspected that the way people had felt, not economic conditions, had been the major cause of the war. Not that he'd be able to prove it.

"I think you're going to find this young man quite interesting," Dr. Larry Edelstein said.

We were walking across the medical school's campus to the psychiatric clinic. The sky was a soft, luminous blue. Sunlight sharpened the edges of the low-rise modernist buildings, and the eucalyptus trees hanging over the red brick sidewalk filled the air with the sharp medicinal odor of cough drops.

"His name is Chase Granger," Larry said. "He's seventeen, a high-school junior, and his strategy for coping with his social environment is less than optimal."

Larry and I had been friends since I was in graduate school and he was doing a residency in psychiatry. He was athletic, with bright blue eyes and sandy hair. A cowlick sticking up like wispy feathers from the crown of his head gave him a boyish appearance. "Is it particularly hostile?" I asked.

"Yeah, high school," Larry said. He looked at me without smiling. "I don't mean that as a joke."

"I didn't take it that way," I said. "Does anybody still believe the best-years-of-your-life myth?"

"Certainly not Chase," Larry said. "School-related problems in late adolescence are usually associated with either academic or social factors. For Chase, it's social."

"Did his parents tell you that?"

"I haven't talked to them," Larry said. "But I know from the intake data that he goes to the most exclusive prep school in the area, and he's in the top five percent of his class."

"Five percent?" The figure seemed oddly precise.

"The school prides itself on being academically tough," Larry said. He gave a crooked smile. "It calculates GPAs to the third decimal."

We both knew about such schools. Social and academic pressures would be immense. Recreational drugs would tempt many and hook a few. Among the girls, eating disorders would be frequent, and almost everyone would eventually experience a psychological crisis. Suicidal gestures, and sometimes suicide, would occur every year. Kids would whisper about them, but most parents would never hear the details. Although many students would get help, a larger number would need it and not get it.

"A rich kid with problems?" I asked.

"A scholarship boy with problems," Larry said. He shifted his black leather folder from his right hand to his left.

"Like Orwell's miseries at his prep school," I said.

"Exactly," Larry said. "I'm not sure I can help this kid much. He's a cutter, and we don't have much to offer them."

"He must be intelligent," I said. "Otherwise you don't get to be in a top percentile in a competitive school."

"He is, and that means he could be a candidate for a cognitive approach," Larry said. Seeing my puzzled look, he added, "If he understands why he's doing what he's doing, he may be able to change it."

"Is he coming to you voluntarily?"

"Not really," Larry said. "His gym teacher noticed cuts on his thigh and informed the school's psychologist. She called his parents."

"Did anybody talk to Chase?"

"Not clinically speaking," Larry said. "His difficulty is beyond the scope of the psychologist's expertise, but the school's medical insurance covers scholarship students."

"*Mens sana in corpore sano,*" I said.

"It's that kind of place," Larry said. "Chase's parents were freaked out by the cutting and insisted that he come here."

"So he's not a willing patient?"

"That remains to be seen," Larry said. He shrugged. "If he's not willing to work with me, I can't do much."

The second-floor consulting room resembled the lounge of the Yale Club. The walls were paneled in dark wood, and the thick carpet was loden green. Squat leather club chairs were arranged around a circular coffee table with a white marble top. Near the ceiling of the outside wall, a strip of narrow windows let in the clear light of the afternoon sun.

Chase Granger was sitting facing the door. A can of Diet Coke was on the table, and a black and silver backpack was on the floor beside his chair. He stood up when we came into the room, and Larry introduced us.

"I'm glad you came today," Larry said. He put his portfolio on the table. "I see somebody's already fixed you up with something to drink." He pointed at the Diet Coke. "Is that okay, or do you prefer something else?"

"I'm good," Chase said. His voice was tight.

Chase was short and maybe fifteen pounds overweight. His nose was large and had a bump in the middle, as if it had been broken. Tight curls gave his black hair a messy, tangled look, and his hazel eyes were never still. He glanced at me from time to time, but his focus was on Larry. He seemed to be studying Larry, trying to guess his thoughts.

"Sit back down," Larry said. "I'm going to get us something to drink."

Larry opened the refrigerator in the built-in bar at the back of the room. I nodded when he held up a small green bottle of Perrier, and he came back with two of them. Larry took the chair facing Chase, and I sat on his right

"I bet you've been dreading talking to me," Larry said. He wasn't looking at Chase, and he seemed to be concentrating on unscrewing the cap from the Perrier bottle. "Maybe you think you'll be embarrassed or that I'll repeat everything to your parents or to the school."

Chase apparently hadn't expected Larry to be so direct. His eyes showed his surprise, but he kept quiet. He still seemed to be studying Larry. He crossed his arms over his chest, as if he needed to protect himself.

"Am I right?" Larry asked. He still didn't look at Chase.

"I guess so," Chase said. His voice was low and husky.

"I'd feel the same," Larry said. He put his bottle down on the table, then pushed it toward the center. He finally looked up at Chase. "So let me start with two points. First, you can't tell me anything I haven't heard many times. Second, because you're my patient, nothing you say to me will be passed on."

"Not to my parents?" Chase sounded skeptical. "Even though I'm a minor?"

Larry shook his head. "Only if I suspect you might harm somebody." He leaned toward Chase, studying his face. "Have you had thoughts along those lines?"

"God, no." Chase sounded shocked. "I'd never hurt anybody, no way." He stopped abruptly and lowered his gaze. "I did fantasize about shooting this guy Lopez who kind of humiliated me." He shook his head. "But I'd never, ever do it"

"This was at school?" Larry asked.

"Sure," Chase said. "Everything in my life is about school." He gave a long sigh. "Everything good, everything bad."

"What did Lopez do?" Larry asked.

Chase hesitated. He seemed wary, as if considering how much he wanted to reveal. Larry remained quiet, but he continued to look expectant. Rather than prying, he simply waited.

Chase uncrossed his arms and took a sip of his Diet Coke. He set the can back on the table, taking great care to center it on the square blue napkin under it. Placing his hands flat on his knees, he finally began to talk. He told about what his life was like at College School, then he described the incident with Lopez. He spoke in a matter-of-fact way, but his voice was strained. Occasionally, it broke, and he had to clear his throat.

"I sliced up my leg the same day," Chase said. His wariness had eased while he was talking, but he continued to study Larry's face for a reaction. "Right after school."

"Did it make you feel better?" Larry asked. He gave no sign of approving or disapproving. He remained only curious.

"Yeah," Chase said. "It did." He seemed surprised Larry could imagine that it might. "I felt guilty later, but that first cut is like taking a powerful drug."

"That's quite perceptive of you," Larry said. He smiled and nodded, as if Chase had solved a difficult problem. "If you don't mind telling me, when did you start cutting yourself?"

"Last year," Chase said. "Maybe March. I was feeling bummed about school. But my parents are so proud that I go to CS I can't tell them how hard it is for me sometimes."

"To protect them?" Larry asked.

"Something like that," Chase said. "They work hard, and neither went to college. Charlie, my little brother, plays these dumb computer games and is a total slacker." He shrugged. "So that leaves me, and I don't want to disappoint them."

"Your grades are outstanding," Larry said. "Is that about pleasing your parents?"

"Maybe that's how it started," Chase said. "But I like learning things and making good grades." He took another sip of Diet Coke and put the can back on the napkin without trying to center it. "It lets me think I'm not totally pathetic."

"When you cut yourself, do you ever think of doing anything more drastic?" Larry asked. His smile had disappeared, and he sounded earnest. "You mentioned you imagined shooting Lopez."

"Do I think about killing myself?" Chase asked. "Never, absolutely not." He smiled, looking embarrassed. "I dip the knife blade in alcohol so the cuts won't get infected."

"You never think, 'I'm so miserable that I want to end it permanently'?" Larry's voice was even, but it had become harsher. He sounded almost accusatory.

"No, not once," Chase said. He wriggled around in the big chair, suddenly agitated. "It would kill both my parents. My dad flat-out couldn't handle it. He's like forever telling me how proud he is of me and how I'm going to get a scholarship to Berkeley or Stanford."

"That's a lot of pressure," Larry said. His tone again became conversational. "Do you think that's why you cut yourself?"

"It's not my family," Chase said. "And it's not even Lopez." Shaking his head, he closed his eyes and frowned. "It's hard to explain, but when I'm at school and not in class, I sort of turn off. It's like I become a zombie."

"So why the cutting?" Larry asked.

"For the thrill," Chase said. He paused to check Larry's response, then he continued. "Last summer, I worked at the branch library in Hilldale, and I didn't cut myself more than twice."

"Do you know why you cut back?" Larry caught himself, then laughed. "Sorry, Chase" he said. "Poor word choice."

Chase also laughed. His wariness had disappeared, but talking about himself seemed to unsettle him. He squirmed around more in the club chair, unable to find a comfortable position.

"I think it was because I was busy checking out books and helping people do research," Chase said. "I was with patrons and staff all the time, so I didn't feel so isolated." He paused, then smiled at Larry. " 'Cut off from people,' you might say."

Larry returned the smile, but he made no comment. He seemed distracted, as if he were thinking about something else. Then he slid forward in his chair and leaned toward Chase.

"From talking to you and knowing your school record, I think you're quite intelligent," Larry said. "You also seem to have a lot of insight into why you cut yourself."

Chase was paying rapt attention.

Larry held up two fingers in a V. "Those two factors help me understand you. But they may also help you decide how to deal with your behavior." He sat back in his chair.

"I don't just do what you tell me?" Chase asked. He sounded surprised, rather than resentful.

"It's your choice," Larry said. "If you had an inflamed appendix, you'd have to decide if you wanted a surgeon to operate. End of story." He held up his hands with the palms out. "In psychotherapy, if you decide to change your behavior, I can't put you to sleep and take it from there. You have to be an active participant in an often uncomfortable and frustrating process."

"I think I get it," Chase said. He rubbed a hand over his mouth. He was beginning to look tired.

"I'm sure you do," Larry said. "I'm going to tell you a few things about people who cut themselves. I think you'll be able to use the information, so the sooner you have it, the better."

"Should I write anything down?" Chase asked. He glanced at his backpack on the floor.

"You'll remember," Larry said. "But if you want to take notes, that's fine." He took a sip from the green bottle.

Chase unzipped his backpack and took out a spiral notebook and a mechanical pencil. He opened the notebook, then rested it on his left knee. He glanced at Larry, ready to listen.

"Three take-home points," Larry said. "First, cutters use physical pain to distract themselves from emotional pain." He kept his eyes on Chase. "Because they can control the physical pain, they can make the emotional pain go away for a while."

Chase's face revealed no surprise, but he listened to what Larry was saying with intense concentration. Larry paused, and Chase bent over and wrote something in the notebook.

"Point two, cutting isn't a cry for help," Larry said. "But it is an expression of distress, and some cutters can become suicidal." He dropped the formal tone. "That's why I asked if you'd ever thought about ending your life."

"It's not just because I don't want to hurt my parents," Chase said. He spoke earnestly, as if welcoming another chance to explain himself. "Down deep I know I'm not going to be in school forever and that things will get better later."

"Exactly right," Larry said. He gave Chase a reassuring smile. "My last point you already hinted at. You said the first cut is like taking a drug."

"Right," Chase said. He continued to give Larry his complete attention. "It produces instant relief, the way you said."

"Cutters count on that response," Larry said. "Each time they cut themselves and feel better, that reinforces their behavior and makes it more likely they'll cut themselves again."

"So they become addicted." Chase nodded, then bent his head to make another note.

"Whether it's true addiction is debatable," Larry said. "But the relief cutting produces turns it into a habit." Larry looked at the ceiling a moment, then back at Chase. "It's like discovering that having a shot of vodka makes you feel better."

"I get what you mean," Chase said. "So even people who aren't alcoholics start looking forward to a drink."

"Exactly," Larry said. He sounded like a teacher pleased with a student's answer. "That's why it's so hard for cutters to change their behavior."

Chase glanced up from his notebook with a puzzled expression.

"Think about the addictive aspect," Larry said. "Cutters are in emotional pain. They cut themselves, they feel better." He raised an index finger. "They have discovered a mechanism for coping with their unhappiness."

"So they don't want to give it up," Chase said, nodding. He made another entry in his notebook. When he lifted his head, he looked across the table at Larry.

Larry remained silent, waiting for Chase to speak. The sun was low enough to shine through the strip of high windows and spill across the ceiling. Light passing through the two Perrier bottles threw shadowy green patches on the white tabletop.

"Dr. Edelstein," Chases said. He drew out the name, then paused. Finally, speaking quickly, he said, "I'm sure this isn't what you want to hear, but I'm one of those who doesn't want to give it up." He wiped the fingertips of his right hand over his lips. "Cutting, I mean."

"I'd be surprised if you did," Larry said. He smiled. "People don't like to quit doing the one thing they can rely on to make them feel better."

"But didn't you say I got to choose?" Chase asked. He narrowed his eyes and gave Larry a suspicious look.

"You do," Larry said. "But you have to think things through before making a choice. You need the cutting because of emotional pain, so what does that suggest to you?"

"That if I didn't have the pain, I wouldn't need the cutting?" Chase's answer was reluctant, as if he were being pulled in a direction he didn't want to go.

"Just so," Larry said. "Some cutters continue the habit to fight boredom, but I don't think boredom is a problem for you."

Chase shook his head.

"The question about you is whether you can get to feel better," Larry said. He gave Chase an appraising look. "If so, that will reduce your need to cause yourself pain."

"Maybe if I go to Vegas and win five million," Chase said in a sardonic tone. "Then everybody would want to be my friend."

"Then you'd be here talking about another set of problems," Larry said. He nodded toward the leather folder on the coffee table. "The diagnostic tests show that you're a little depressed. Not crawl-under-the-bed-and-hide depressed, but enough to reinforce your feelings of social isolation."

"I have felt kind of down," Chase said. "I figured it was due to the way my life is right now."

"That's a contributing factor," Larry said. He took a sip of Perrier, then stifled a burp with the back of his hand. "But even now, your life doesn't have to seem so bleak. I can work with you on that."

"You mean, like drugs?" Chase said. He raised his eyebrows.

"An antidepressant might be useful," Larry said. "But a drug won't make the bad stuff go away. Antidepressants aren't like antibiotics. We need to meet regularly and talk about the things that bother you most and develop some strategies to help you deal with them."

"You make it sound so practical," Chase said. He gave a small smile. "I thought psychiatrists made you talk about your dreams and the terrible things that happened in your childhood."

"Treatment is practical," Larry said. "That's not to say it's easy." He smoothed down his cowlick, as if it suddenly bothered him. "And you don't need to be Freud to see that dreams and childhood experiences can sometimes help you understand your present condition."

"So what's the next step?" Chase clipped his pencil to his shirt pocket.

"I'm going to write you a prescription and ask you to make an appointment with me for next week," Larry said. "Whether you take the medicine or keep the appointment is up to you."

"Not up to my parents?" Chase gave Larry a skeptical look.

"Up to you," Larry repeated "You're intellectually mature enough to decide if you want me to treat you."

"I definitely do," Chase said. "I thought this was going to be a horrible experience, but I'm glad I came."

"Then I'm also glad," Larry said. He smiled at Chase. "I have one more question."

Chase sat up straight, looking worried.

"Did the geometrical figure you cut on your leg have any particular significance?" Larry asked.

"Not really," Chase said. He gave a short laugh and relaxed again. "It's a pentagram. Last year in World Literature, we read *Faust*, and our book showed a picture of it. Faust draws it on the floor to control the devil." He shrugged. "I liked the design, but it doesn't have any special meaning for me."

"Oh, really?" Larry said. "Maybe you should give the topic some more thought," He stood up and held out his hand to Chase. "Something interesting may occur to you."

Chase is seventeen, which makes him a minor.

Thus, even if he has as much money as Richie Rich, the Richest Kid in the World, he can't buy Trump Tower, order a gold-plated Lexus, or turn over half his fortune to Bono to save the world.

State laws define the age of majority, and it's never lower than eighteen. Minors lack the legal status for entering into contracts or, except in special circumstances, making decisions about their medical care.

MEDICAL CARE

If Chase stumbles, cracks his head, and develops a blood clot that presses on his brain (a subdural hematoma), he can be treated immediately in the ER. The ER doctors can rely on the notion of parental *presumed consent*: that his parents, if present, would consent to Chase's being treated to prevent brain damage.

But suppose Chase consults Dr. Deborah Handy, a plastic surgeon, and asks her to reshape his nose. Even if he has the cash in hand, Dr. Handy can't operate. She must talk to Chase's parents and explain the surgery, list its risks and benefits, and mention alternatives (such as doing nothing). Although the nose is Chase's, his parents must give their *informed consent* for the surgery.

Dr. Hardy must have a similar discussion with Chase, because she also needs his *informed assent*. The assumption in ethics and law is that minors aren't capable of exercising mature judgment. Thus, they aren't able to give informed consent. If given an explanation of what is going to be done to them and why, however, they can give informed assent.

CONFLICT

If Chase wants Dr. Handy to alter his nose and his parents agree, the surgery can proceed. Yet if Chase assents but his parents refuse to consent, the surgery can't happen. Nor can it happen if Chase's parents want his nose altered and Chase refuses to assent. The situation is different, though, when a minor has a serious disease requiring treatment. If the parents give consent, but the adolescent refuses to assent, the hospital usually goes to court and asks a judge to order them to treat the child.

Courts, in a few sad cases, have allowed adolescents to decide to discontinue chemotherapy against the wishes of their parents. These have been cases in which earlier treatments failed and additional ones seemed unlikely to succeed.

MENTAL ILLNESS

Mental illness occupies a special status with respect to informed consent. Chase was sent to see Dr. Edelstein by his parents, but he could have gone on his own, and he doesn't need their consent to be treated. In his state, as in most, minors can seek treatment for mental health problems without involving their parents.

The reasoning behind this exception to the parental informed-consent requirement is that people in serious mental distress may cause harm to themselves or others. If they need to go to their parents and ask for consent, they might avoid treatment or their parents might not realize the importance of getting immediate help. Hence, more damage, in the form of personal suffering, injuries, suicide, and homicide, might be done by requiring parental consent than by waiving it.

CONFIDENTIALITY

Dr. Edelstein assures Chase he won't reveal Chase's secrets to his parents or his school. This is the first step in the standard psychiatric practice of discussing with adolescent patients exactly what information they want withheld from their parents. A psychiatrist has an obligation to protect his patients from the harm they might suffer if he reveals certain facts about them. If a patient has been lying to her parents about getting an abortion and they find this out, for example, they might turn her out of the house, assault her, or even kill her.

Yet psychiatrists can't guarantee they will never violate patient confidentiality. Dr. Edelstein questions Chase closely to determine if Chase is suicidal. He also makes a deliberate effort to discover whether Chase's fantasy about pulling a gun on Lopez constitutes a homicidal threat. If Dr. Edelstein had decided Chase was suicidal or homicidal, he would have had an obligation to try to prevent Chase from killing himself or Lopez.

That Dr. Edelstein has a duty to keep a patient from killing himself is obvious. The patient is suffering from a mental problem and comes to him for help. Helping a suicidal patient stay alive so that his underlying difficulty can be addressed is a part of treating the patient appropriately.

Less obvious is whether, if Chase tells Dr. Edelstein he intends to shoot Lopez, Dr. Edelstein has a duty to tell Lopez. Since a 1976 ruling by the California Supreme Court, many states have passed laws requiring a psychotherapist whose patient threatens to harm someone to warn that person. What prompted the court decision is explained by what happened to Tatiana Tarasoff.

THE TARASOFF CASE

In 1968 Prosenjit Poddar and Tatiana Tarasoff were students at the University of California, Berkeley. Poddar was studying on a visa from India, and the two met at a folk-dancing class. At a New Year's Eve party welcoming in 1969, Tarasoff kissed Poddar. This convinced him that she had passionate feelings for him, but the next time he talked to her, she denied having any interest in him. She told him she was seeing other men and didn't want to have anything more to do with him.

Poddar became severely depressed. He neglected his classes, suffered fits of weeping, and sometimes made no sense when he talked. When he told a friend that he was thinking about blowing up Tatiana's room, the friend convinced him to go to the Berkeley student health service.

Poddar became a client of Dr. Lawrence Moore, a staff psychologist. In August 1969, Poddar, during his ninth therapy session, told Dr. Moore that he was planning to kill Tatiana when she returned to Berkeley to start the fall semester.

Dr. Moore reported the threat to the campus police. He told them he thought Poddar was dangerous and should be hospitalized, involuntarily if necessary. The police picked up Poddar and questioned him. They soon released him, however, because, according to the police report, he convinced them that he had "changed his attitude" toward Tatiana. Poddar promised the police that, when she came back to Berkeley, he would stay away from her.

Dr. Harvey Powelson, the director of the student health center, was briefed about Poddar's threat and the police decision. He decided the center shouldn't make any additional effort to have Poddar hospitalized. No one informed Tarasoff or her parents that Poddar had threatened to kill her.

Poddar stopped showing up for his therapy sessions. On October 27, 1969, he went to the house where Tarasoff lived. Knowing of no reason to fear him, she invited him to come inside. He then stabbed her to death with a kitchen knife.

Poddar was convicted of second-degree murder, but his conviction was overturned, on the grounds that the jury had been improperly instructed. Prosecutors decided against a second trial, and Poddar was released and deported.

DUTY TO WARN

The *Tarasoff* opinion held that a therapist who believes a patient poses a danger to someone has a duty to use "reasonable care" to protect the intended victim. This may involve warning the person, but if it's reasonable to believe a warning might be insufficient, the therapist has a duty to take steps to have the patient involuntarily institutionalized.

Dr. Edelstein decided that Chase posed no danger to Lopez, but he could have been wrong. Many therapists object to laws that assign them a duty to warn potential victims, because therapists aren't good at making predictions about whether a patient is dangerous.

In a study cited in the Tarasoff case, psychiatrists rated 989 patients as dangerous and recommended that they be confined in maximum-security hospitals. Yet, for legal reasons, all the patients were sent to ordinary hospitals, and about two hundred of them were discharged after a year. Within that year, only 7 of the 989 had done anything, or threatened to do anything, requiring them to be confined in a maximum-security facility. So, professional groups representing therapists argued, if therapists are so poor at predicting that a patient is going to be dangerous, why should they have a duty to warn when a patient threatens to harm someone?

The answer in the *Tarasoff* decision can be put succinctly: Better safe than sorry. The court acknowledged there would be many false alarms, but it decided that if a therapist "does in fact determine" or "reasonably should have determined" that "a patient poses a serious danger of violence to others, he bears a duty to exercise reasonable care to protect the foreseeable victim of the danger."

Cutting can be a symptom of a serious psychiatric disorder. However, those who use sharp blades to score their skin are rarely suicidal and hardly ever homicidal. Thus, Dr. Edelstein isn't likely to need to violate Chase's confidentiality. The real question, as he tells Chase, is whether he will be able to help him give up his habit. Cutters lack the equivalent of methadone pills or the nicotine patch to smooth over the rough spots. They have to...just do it!

TEN

It Seemed Like a Good Idea

MEG ANDERSON DIDN'T look like someone with a fatal disease.

She was tall and slender, with an athletic, almost boyish build. Her hazel eyes were luminous, and her hair had the golden-dark color of wild honey. It was so short that feathery tufts stuck up along the whorl at the crown of her head, but the result was charming. She was like a beautiful actress pretending to be seriously ill for the sake of the story. Only Meg wasn't pretending, and no one knew how her story would end.

The year was 1995, and Meg was thirty-five years old. She had been diagnosed with breast cancer a few months earlier, and her hair was so short because it was still growing back after chemotherapy. Only 1 in 2,300 women her age develops breast cancer, so she was statistically unlucky.

Nor did her bad luck end there. When Meg was diagnosed, her doctors discovered that her cancer had already progressed to a late stage. Cancerous cells had migrated to her lungs, and she was judged to have the most advanced state of breast cancer: stage IV. Women with stage IV cancer usually have been treated and become cancer free, then their disease returns months or years later. Thus, Meg started off where most women diagnosed with breast cancer fear they might end up.

I met Meg for the first time when I visited her hospital room with Dr. Timothy Terrence. She had gone through a standard course of treatment in Cleveland, her hometown. Her cancer hadn't been driven into remission, so she had come to Southeastern Medical

Center to consider an experimental treatment that might help her. Dr. Terrence, an oncologist-hematologist, was the physician responsible for her care.

Meg was on the fourth floor in a large room with yellow walls. She had her own bracket-mounted TV and a private bathroom, but the room had only one window. It presented a dreary view of the circular driveway in front of the hospital. The bright spot in the room was a mixture of red, pink, and yellow lilies in a glass vase on the low dressing table. Meg was sitting in a chair beside the flowers when we came in.

"I'm happy to see somebody," Meg said. She stood up and gave each of us smile as she shook hands. "I get so bored that I'm almost glad when the nurses come in to draw blood."

She was wearing a faded pink hospital gown and a matching wraparound bathrobe. Her sparse makeup emphasized the greenish hue of her hazel eyes. Her skin was pale, but the faint rosy blush high on her cheeks seemed natural.

"If you're bored, you can't be feeling too terrible," Dr. Terrence said. He was tall and reserved, with an easy manner and an unhurried air.

"My chest is still sore from where the chemo port was sewn in," Meg said. She put the flat of her right hand on an area above her collarbone. She then sat beside the dressing table, with her back to the hospital bed and the wires, hoses, and read-outs of the monitoring equipment.

Dr. Terrence sat facing Meg, and I took the chair beside the window. The shade was halfway up, and I had a clear view of the awkward press of cars jockeying around the entrance.

"The area didn't look infected yesterday," Dr. Terrence said. "Give it a couple of more days, and it'll be better."

"That's good to hear," Meg said. She smiled, and a brightness came into her eyes. She gestured toward the vase. "Do you like my flowers?"

"Lovely," Dr. Terrence said.

"My friends Amanda and Heather sent them." Meg gazed at the flowers and shook her head gently. "I'm so lucky."

Timothy Terrence and I were collaborating on a book about experimental therapies. Meg would have to decide whether to have a so-called bone-marrow transplant, a treatment not tested in clinical trials. That made her an appropriate case for us, and she had agreed to be interviewed. Dr. Terrence would make sure she had all the facts needed to make an informed choice, and if she chose the treatment, he would oversee it. I was present only to observe.

"I want to hear about your problems from the beginning," Dr. Terrence said. He held up a thick manila folder. "I've read your chart, but it's my rule to listen to the patient instead of relying exclusively on what the doctors write."

"You must be a radical," Meg said. She gave him an ironic smile. "My surgeon didn't believe in wasting time on conversation."

"People get in a hurry," Dr. Terrence said. "I've got plenty to tell you, but right now, I want you to do the talking."

"I've always liked being the center of attention," Meg said. Her tone was playful. Then her expression became serious, and she lowered her eyes. "I'm not exactly sure what you need to know, but I'll tell you the whole story."

Meg majored in history in Cleveland State, and even though she had known little about computers, when she graduated, she got a sales job with a software-development company. She stayed with the company, and as it grew, she became the supervisor of a sales group. She lived in a two-bedroom apartment about a mile from Lake Erie. She knew people at work, but her closest friends, Amanda and Heather, were young, single women living in her building. She worked hard, yet she found time for the social life of impromptu dinners, evenings at the movies, and small parties typical of people her age.

Steve, Meg's boyfriend, was an accountant, but she considered him too dull and egocentric to marry or live with. But she did like having someone to go out with, so she got cold feet every time she thought about dumping him. Maybe she also kept him around, she speculated, because she was starting to worry about aging. She could already see small dimples of cellulite on her thighs, and that was depressing.

"I noticed a patch of skin on my left breast that looked like cellulite," she told Dr. Terrence. She gave a small laugh. "I thought, well, you're thirty-five, so what do you expect?"

She wasn't worried about the dimpled skin, so she didn't see a doctor until several months later, when she went for her annual gynecological checkup.

"Is there a cure for cellulite?" Meg asked Dr. Florence Long, her gynecologist. Meg pointed to the place on her breast. "I'd like to get rid of this patch."

Dr. Long examined the patch and immediately recognized its significance. Such dimpling is called a *peu d'orange* (orange peel) sign. The orange-peel texture is caused by cancer cells tugging on the skin from below. By the time the sign appears, the tumor has grown beyond its early stages.

"Dr. Long told me I needed a mammogram immediately," Meg said. "While I was getting dressed, she phoned a radiologist in the same building who said I should come right down."

Dr. Long called Meg at home that evening to tell her that the mammogram revealed that her left breast had a three-centimeter mass directly below the dimpled skin. Her right breast seemed normal. The lump, about the size of a walnut, would have to be biopsied, and Dr. Long had already arranged for Meg to see Dr. Susan Bronstein, a surgical oncologist.

Meg thought she'd been prepared to hear bad news, yet she was surprised and stunned. "No one on either side of my family ever had breast cancer," she recalled. "I always knew it could happen to me, but I figured that I'd be in my sixties or seventies." She gave a wan smile. "Not my thirties."

BIOPSY

Dr. Bronstein, a slim, black-haired woman hardly older than Meg, was brisk and businesslike. Meg perched on the edge of the examining table while Dr. Bronstein spent several minutes examining both of Meg's breasts. She then ran her fingers around Meg's armpits and along her shoulders. She was searching for swollen lymph nodes, but she found none.

Dr. Bronstein snapped the film of Meg's left breast onto a lightbox and studied it for a long time.

"I'm going to give you some numbing medicine, then do the biopsy with another needle," Dr. Bronstein said. She uncapped a plastic syringe and injected an anesthetic into the dimpled patch on Meg's left breast.

After giving the anesthetic time to work, Dr. Bronstein inserted a large needle into the center of the tumor. She pulled up on the plunger, then withdrew the needle. A drop of blood seeped from the puncture. She wiped off the blood and put a Band-Aid over the small wound.

"Please get dressed," Dr. Bronstein said. She was so businesslike that Meg thought she was like a machine. "Did you bring anyone with you?"

"No," Meg said. "My parents live in the D.C. area, and I didn't want to bother my friends." She felt the need to explain, so Dr. Bonstein wouldn't think she was a pathetic outcast. She didn't like her surgeon, but it seemed important to make a good impression on her. "I didn't want my boyfriend here."

Dr. Bronstein nodded and said, "We'll talk in my office."

Meg sat on the beige sofa in the seating area at the front of the spacious office. She was sorry she hadn't asked Amanda or Heather to come with her. She felt very alone.

Dr. Bronstein sat on the boxy chair opposite Meg and looked directly at her. "I can't know for sure until I get back the biopsy results," Dr. Bronstein said. "But my examination and the mammogram both point to cancer in your left breast."

"I was afraid of that," Meg said. She felt abstracted, as if she were sitting in a movie theater listening to dialogue that concerned only the characters on the screen.

"When breast cancer has spread to the skin, it's also probably spread to other places," Dr. Bronstein said. She sounded like a textbook. "If the biopsy confirms my suspicion, I recommend we start with a lumpectomy, followed by chemotherapy. I can perform the surgery, but I'll refer you to Dr. Alan Soames for the chemotherapy." Speaking more rapidly, she then added, "You're free to get a second opinion."

Meg shook her head. She had heard the same diagnosis from Dr. Long, the radiologist, and now Dr. Bronstein. Unless the biopsy results turned out to be a surprise, she saw no point in seeing more doctors.

"When can you operate?" Meg asked.

She would have had the surgery that afternoon, if possible. She wanted the cancer out of her. For the first time in her life, she felt betrayed by her body. She'd been careful with her diet since college, and she had been physically active forever. Two days ago at the office somebody had offered her a doughnut, and she'd turned it down. Now she wished she'd eaten it.

"I'll want you to get some X-rays and a bone scan, if the biopsy is positive," Dr. Bronstein said. "We can schedule surgery then, but let's not get ahead of ourselves." She paused for a long moment, then said, "I'm very sorry."

The expression of sympathy was awkward and impersonal, the words sounding more mechanical than genuine. Even so, Meg felt tears stinging the corners of her eyes. A heaviness settled over her, and she suddenly felt sorry for herself.

She was a woman with breast cancer.

When she got home, Meg called Amanda and Heather and invited them to her apartment for dessert. After the frozen cheesecake and coffee and the usual chatter about work and boyfriends, she told them about her diagnosis. Both were upset and more ready than she was to deny the facts.

"You don't *know* you have breast cancer," Amanda said. "The biopsy may show it's something else."

"You need talk to more doctors," Heather said. "Go to one of the big medical centers, like the Cleveland Clinic."

They talked for almost two hours, and Meg had a hard time persuading her friends to leave so she could get to bed. She was exhausted, and she was having a hard enough time holding herself together without spending energy reassuring them that she would be all right. Particularly since she wasn't sure she would.

She thought about calling her parents and Steve, then decided she wasn't up to it. Her parents would need more reassurance than her friends, and she didn't know how Steve would react. Cancer or not, she had to meet with her sales group in the morning, so she needed to rest.

The next day, late in the afternoon, Dr, Bronstein called Meg at work. As a sales manager, she rated an office, and she asked Dr. Bronstein to wait while she closed the door.

"Go ahead, please," Meg said. Her hand felt clammy against the plastic handset.

"I've just received the path report," Dr. Bronstein said. "You have a cancer called infiltrating ductal carcinoma."

Meg lost herself in her screen saver for a moment. Pink rosebuds opened into blossoms, then a wind blew away the petals. Swirling in the air, the petals became a swarm of small dots that turned into pink rosebuds. The cycle started over.

"The tumor also tests estrogen-receptor negative," Dr. Bronstein continued.

"Is that good or bad?" Meg asked. She had no idea what Dr. Bronstein was talking about.

"It has treatment consequences," Dr. Bronstein said. "If you were positive, we could give you a drug to block the estrogen receptors, and that would slow down tumor growth."

"I see," Meg said. She knew she didn't really understand the explanation, but the point seemed to be that a drug that might have helped wouldn't work.

"I'm going to give you the number of the hospital's imaging center," Dr. Bronstein said. "I'll call and let them know what X-rays and scans I need."

Meg made her appointment after she said goodbye.

MEG'S CANCER

Meg's cancer, she learned later, had started in the cells lining the milk ducts of her left breast. This made it a ductal carcinoma, and twenty percent of the time, such cancerous cells remain in place. This form of the disease is called "ductal carcinoma *in situ*," and it is usually curable by surgery alone.

Meg's cancer was identified in the pathology report as *infiltrating* ductal carcinoma. This is a menacing form of the disease in which cancerous cells form a tumor and began escaping into the bloodstream. Ordinarily, large numbers of these cells are swept into the lymphatic system and end up in the lymph nodes, causing them to swell. Meg's lymph nodes were normal, but cancer cells may also be carried by the blood directly to the liver, lungs, brain, or bones. Because these cells continue to multiply in their new locations, they can form new tumors and destroy the organs they have invaded.

Meg later discovered that, although Dr. Bronstein hadn't stressed the point, the lack of estrogen receptors on the cells of her tumor was another piece of bad news. The number of hormone receptors on the cells is an indicator of how malignant the tumor is—the more receptors, the less malignant. Also, as Dr. Bronstein mentioned, if a tumor has estrogen receptors, doses of an anti-estrogen hormone can block them. This won't cure the cancer, but by keeping the cells from dividing, the hormone can bring it under control.

ANOTHER PHONE CALL

Meg spent most of Wednesday, the day after she had made her appointment, having breast X-rays, a CT scan of her body, and a bone scan. Because everything had been moving at such a rapid pace, she expected to learn the results at once. Thus, Thursday and Friday were made anxious by waiting for the phone to ring, but when Dr. Bronstein didn't call, Meg figured nothing would happen over the weekend.

She couldn't relax, though, until she'd had the conversation with her parents that she had been dreading. She found it hard to tell them she had cancer, but they were both so comforting and so quick to offer help that by the time she said goodbye, she felt much better and was glad she had talked to them.

Steve called on Friday, but she put off meeting him, and on Saturday night she went to see *Apollo 13* with Amanda and Heather. Except for asking how she was feeling, both of them stayed away from the topic of cancer. She liked the movie, maybe because it was a story about overcoming difficulties by not giving up. More likely, she decided, it was because she'd always had a weakness for Kevin Bacon.

Dr. Bronstein finally called on Monday night around eight o'clock. It was long past the time Meg had expected any doctor to be working. "The radiologist sees no indication of any breast tumors other than the one we know about," Dr. Bronstein said.

"That's good," Meg said. She got the impression from the way Dr. Bronstein hesitated that another shoe was about to drop.

"The chest X-rays show a suspicious spot in the lower lobe of your left lung," Dr. Bronstein said. She paused, perhaps to see how Meg was going to react.

"Is it cancer?" Meg asked. She felt steady and in control, but the light above her kitchen table blurred for a moment. "Is that what suspicious means?"

"It could be cancer," Dr. Bronstein said. She sounded as dispassionate as an expert witness testifying in court. "I don't want to encourage false optimism. The CT scan revealed what the radiologist believes is a probable tumor in your lung."

"That's not good news," Meg said. Her voice was flat. She felt as if she were talking about somebody else. Yet her heart was racing, and her mouth was dry and sticky.

"There is some good news," Dr. Bronstein said. She didn't sound upbeat, though. "The bone scan appears negative. No bright spots on the images, which means the disease hasn't spread to your bones."

Meg felt good about that. She was aware that it was odd to be glad not to have cancer in your bones, after you've just heard you have it in your breast and lung. A few days before, she'd have ridiculed her own feelings, but she was beginning to sense the importance of scraps of encouraging news.

"I see no reason to change my treatment plan," Dr. Bronstein said. "Lumpectomy, then chemotherapy. At the end of the chemo, I'll go back in and take out more tissue from the tumor area. Have you seen Dr. Soames yet?"

"Tomorrow afternoon," Meg said.

"Excellent," Dr. Bronstein said. "Call my office and schedule the lumpectomy. You should avoid any unnecessary delay."

CONSULTATION

Dr. Soames was the opposite of Dr. Bronstein. He was older and less elegant. His hair was streaked with gray, and he was a good twenty pounds overweight. His blue trousers were creased across his lap, as if he spent a lot of time sitting.

"I'm so sorry about your problem," he told Meg. He had already read her file when she met him in his office. "You're going be in big fight, but you won't be in it alone. I'll be there to help you every way I can."

"Thank you," Meg said. Dr. Soames' warmth caught her by surprise, and she squeezed her lips together to fight back tears. She hadn't realized she was so emotionally fragile.

"Dr. Bronstein's strategy is appropriate," Dr. Soames said. "Once she's done the lumpectomy, I'll start you on chemotherapy."

"Is it as horrible as everybody says?" Meg asked.

"It's never easy," Dr. Soames said. He smiled. "But it's better than it used to be. We've got good drugs to fight the cancer, and we're better at dealing with the side effects."

"Will I lose my hair?" Meg was embarrassed to ask. It seemed such a petty concern when her life was stake.

"Some people don't, but that's rare," Dr. Soames said. "But it will almost certainly grow back." He hesitated, then frowned, and his eyes looked sad. "Your periods will stop, and they won't start again. You'll go into premature menopause."

"So I can never have kids?" Meg asked.

"I'm afraid not," Dr. Soames said.

She had always expected to be a mother, but at the moment, she felt little regret. She was more focused on herself, which was probably why she'd asked about losing her hair.

They talked for another ten minutes about the drugs that Dr. Soames would use and what Meg should expect during the course of the treatment. The nausea, vomiting, and mouth sores would be unpleasant, but she could get through them. The side effect likely to bother her most, she decided, was the fatigue, because she couldn't afford to quit her job or even take much time off.

SURGERY

At six o'clock on a bright, hot Tuesday morning, Amanda drove Meg to the hospital's Outpatient Surgery Center. After Meg checked in and a plastic bracelet was attached to her wrist, she was escorted to a cubicle and instructed to change into a blue hospital gown and lie on the wheeled bed that took up most of the floor space. Amanda hung up Meg's clothes in the locker in the corner.

"You don't have to stay here," Meg said. She twisted her head right so she could see Amanda.

Amanda was a graphic designer for a large printing company and worked flexible hours. She could put in almost a full day before she'd have to return to pick up Meg.

"I don't want you to feel abandoned," Amanda said. She picked up Meg's left hand and squeezed it.

"You'll be taking me home," Meg said. She returned the squeeze and smiled. "I don't feel neglected."

"Well, then, I'll be seeing you later," Amanda said. She bent over and kissed Meg's forehead. "I'll tell them at the desk to call me if the plans change."

Meg felt a moment of regret as Amanda was leaving. She had little time to feel alone, though. Julia, a tall, gray-haired nurse came into the cubicle and introduced herself. She started an IV, then asked Meg a series of questions about her present health and medical history.

Julia had barely left when a slim Asian man carrying a clipboard came in and introduced himself as Dr. Lee. "I'm going to be your anesthesiologist," he said. He put his stethoscope or her chest, then on her back. "Your lungs are clear," he said. "Are you allergic to any drugs?"

Dr. Lee went on to ask many of the same questions Julia had asked. They were also questions she'd already answered on the registration form. Meg was about to complain, then she realized that having the same questions asked several times was the sort of redundancy you ought to have in a system involving life-and-death decisions.

Meg spent the next half hour watching nurses, doctors, and patients pass the opening of her cubicle. Then Dr. Bronstein arrived. "Doing okay?" she asked, squeezing Meg's wrist. She was in pressed green scrubs and wore no makeup. Her short black hair was pushed away from her face, and she looked both glamorous and competent. "I also want to put in a catheter for chemotherapy today. But I'll do that only if everything else goes smoothly."

"I hope it does," Meg said. "I'd rather not do this again."

"You'll be fine," Dr. Bronstein said. She gave one of her smiles that seemed more calculated than genuine. Then she was gone. Meg wondered, not for the first time, if she shouldn't have talked to another surgeon.

On Thursday, two days after the surgery, Meg was back in Dr. Bronstein's exam room. She was wearing the backless green paper gown the nurse had given her and felt vulnerable and absurd.

Dr. Bronstein tapped on the door, then stepped inside. "Good to see you," she said. She gave Meg one of her formal smiles, then picked up a file from the metal rack on the back of the door. She was silent a moment as she read the first page.

"Let's see how things look," Dr. Bronstein said. She put the file into the rack and pulled down Meg's gown. "When I removed the tumor, I took out enough tissue around it to leave clean margins."

"Clean margins means no cancer cells?" Meg asked.

"Right," Dr. Bronstein said. She touched Meg's left breast. "Any pain?"

"Not in my breast," Meg said. "But where you put in the tube is very sore and hurts when I lift my left arm."

"That's not unusual," Dr. Bronstein said. She pulled away the bandages from the breast and the catheter. "You'll get more comfortable over the next couple of weeks."

Dr. Bronstein was quiet a moment while she conducted her examination. She gave Meg's left breast a careful look, then gently prodded the areas that had been bandaged. Meg stared straight ahead, feeling vaguely detached. She was surprised by how little damage the lumpectomy had done. A thin, raw, red line marked the incision, but the rest of her breast looked normal.

"I see no sign of infection, and the swelling is minimal," Dr. Bronstein said. "When are you starting chemo?" she didn't look up from the notes she was writing.

"Friday," Meg said. "If you say I'm ready."

"You're ready," Dr. Bronstein said. She kept writing. "When you complete chemo, I'm going to want another CT scan to see if the lung tumor has shrunk."

Under her white lab coat, Dr. Bronstein was wearing a dark suit with white piping on the shawl collar. A double strand of dark green jade beads hung around her neck. For a moment, Meg hated her for being well-dressed, confident, and healthy. Dr. Bronstein seemed invulnerable, somebody who would never be foolish enough to get breast cancer.

CHEMO

Meg met Dr. Soames in his office at the hospital. She'd had to wait for thirty minutes and felt a bit put out. He seemed glad to see her, though, and her annoyance vanished.

"I hope you agree with Dr. Bronstein and think the surgery went well," he said. He gave her a sly smile. "Surgeons and patients don't always see eye to eye."

"The catheter hurt more than the lumpectomy," Meg said.

"Even if it's a bit uncomfortable, it's worth it," Dr. Soames said. He was wearing gray trousers, but as he sat down at his desk, Meg noticed that, like the blue ones, these were also creased across the thighs. "We can inject drugs into it, so you don't have to be stuck with needles. These drugs are hard on tissues, so it's best to put them directly into the veins."

"Where do I have this done?" Meg asked.

"In the chemo suite downstairs," Dr. Soames said. "I'm putting you on a three-week schedule. You'll come here every Friday afternoon, and the chemo nurse will give you the drug combination I think will work best for you."

"What are they?" Meg asked. "Even if the names don't mean anything, I like to know what I'm getting."

"Absolutely," Dr. Soames said. "You'll get CAF, which is a combination of cyclophosphamide, Adriamycin, and five-fluorouracil. They interfere with cell reproduction, so they stop tumors from growing."

Meg wrote down the names as Dr. Soames spelled them. He mentioned that they would listed be on the consent form, but she wanted to check them out on the breast-cancer Web sites.

"Since you majored in history, you might like to know that cyclophosphamide is based on the old World War I mustard gas," Dr. Soames said. "And Adriamycin was identified in a sewage line flowing into the Adriatic during the Second World War."

"I saw the name when I was reading about chemo online," Meg said. She looked up from her notebook. "Isn't it the drug they also call Red Death?"

"Don't let that worry you," Dr. Soames said. "It's the cancer cells it's going to kill."

"I'll be okay," Meg said. "I'm just one of those people who want to know the worst, so they aren't taken be surprise."

"I'll always tell you what to expect," Dr. Soames said. "We'll do a blood count every week to make sure you're tolerating the side effects." He laced his fingers together on the desk. "We want to keep you as healthy as possible. So after the first round of chemo, we'll check your blood count again."

"What's the purpose of that?" Meg asked.

Dr. Soames explained that if the drugs suppressed her bone marrow too much, it would stop producing blood cells and platelets. She would develop anemia and be prone to potentially fatal infections. Without platelets, her blood wouldn't clot, and it might seep through the lining of her intestines into her stool. She could bleed to death without noticing any blood.

"Four cycles is the usual treatment," Dr. Soames said. "We'll monitor you as we go along and see if we need to adjust your drugs. At the end, we'll assess you and see if your cancer has gone into remission."

Chemo turned out not to be as bad as Meg had feared. "The schedule was good," she recalled. "I could leave work at three on Friday, get my chemo, then collapse for the weekend. I had a drug that kept the nausea under control. A few hours after treatment, though, I'd feel like my bones had melted. I was so weak I could hardly sit up in a chair."

Meg would stay in her apartment and keep her suffering to herself, then return to work on Monday. "I'd pretend I'd had a great weekend," she said. "Nobody knew what I was going through. I was afraid that if my boss found out, I'd lose my job."

But the chemo began to take its toll. "My hair started falling out in clumps," she recalled. "I hated the wig I bought, because it was hot and itchy and obviously fake. And I had to look at that damned catheter in my chest every morning when I got dressed. It was a constant reminder that I had cancer, and I found that a hard way to start the day."

Meg finally told her boss that she had breast cancer, but she worked as hard as she ever had. "I told myself, I have a little cancer, but I can deal with it. Looking back, I don't think I took the disease as seriously as I should have, because I so much wanted to live a normal life. I've always been a control freak, and I thought I could manage cancer the way I managed everything else."

Eventually, Meg found it impossible to keep up the pretense that her life was normal. She became so depressed that she didn't want to go to work. She even thought about quitting chemotherapy and letting the cancer take its course. Because she was comfortable with Dr. Soames, she told him how she was feeling.

"Dr. Soames put me on Prozac," she said. "In a couple of weeks, I felt enough better to throw away that stupid wig. When I put on my business suit, I started tying a bandanna around my head. That made me feel more like myself, instead of a sick person with cancer."

ASSESSMENT

Meg went to the hospital imaging center for her second CT scan after completing the fourth cycle of chemotherapy. "I was pleased with myself," she recalled. "I was done with chemo and felt more human. Work was fun again, because I could do it without driving myself to complete exhaustion."

Dr. Bronstein called her two days after the scan, and the news wasn't as good as Meg had hoped to hear. The scan showed that although the mass in her lung had grown smaller, it hadn't gone away.

"We might be looking at nothing but dead cells," Dr. Bronstein told her. "I've got to go in and clean up the margins in your breast, so I can do a lung biopsy at the same time."

"And if it shows cancer?" Meg could hardly bear to ask the question. She wasn't so much scared as weary.

"Then additional treatment may be advisable," Dr. Bronstein said. "But let's not get too far ahead. Call my office and schedule the surgery as soon as you can."

"From that point on, it was almost déjà vu," Meg recalled. "Another Friday morning, same hospital, and Amanda dropped me off and picked me up. The nurse was different, but the anesthesiologist was Dr. Lee. He even remembered me." She smiled and shook her head. "It was like having to watch a really bad movie twice."

No cancer was detected in the breast tissue Dr. Bronstein removed. The lung biopsy showed that the chemo had worked on the lung tumor, and most of it now consisted of dead cells.

Most of it, but not all of it.

Meg still had cancer. The cancer cells in her lung made it likely that other cancer cells were circulating throughout her body. Before long, tumors might develop in other organs.

"I was totally bummed out that after having gone though so much, I still wasn't well," Meg remembered. She gave a laugh that was almost a sob. "I was so depressed that I couldn't believe I was taking an antidepressant."

She shook her head and seemed absent for a moment, as if visiting a distant time. "The one good thing that came out of my having cancer was that I found the courage to get rid of Steve. I sent him an e-mail and told him I still had cancer and that I didn't have time for a social life anymore."

"And how did that work out?" Dr. Terrence asked.

"He said he understood," Meg said. She laughed again. "If I had been expecting him to declare unwavering love and eternal devotion, I'd have been very disappointed." She shrugged. "He was nice enough to wish me luck."

Dr. Soames, Meg's oncologist, read Dr. Bronstein's surgical note and the pathology report on the lung biopsy. Although Meg still had cancer, he saw reasons to be optimistic.

"You're a young woman, and your general medical condition is good," Dr. Soames told her at her scheduled office visit. "If you were old and frail or had some chronic illness, I'd tell you that you should focus on making the most of the life you have left."

Meg knew Dr. Soames was trying to be encouraging, but she was skeptical. She still had a disease that sooner or later would kill her, so how could you put a positive spin on that?

"About twenty-five percent of women treated for stage four cancer are alive five years later," Dr. Soames said. He bit at his lower lip and shook his head. "That may seem a low number, but you've got more going for you than most."

"I'm younger," Meg said. She gave him a puzzled look, wondering where he was headed.

"Yes, and you're not being treated for heart disease, diabetes, high blood pressure, kidney failure...and on and on," Dr. Soames said. He gave her a small smile. "And believe it or not, you had a very good response to chemotherapy."

"Isn't that like, the operation was successful, but the patient died?" Meg asked.

"Not exactly," Dr. Soames said. "You tolerated the drugs, and they killed off most of the tumor cells. I think you're a good candidate for additional treatment." He smiled again, as if informing her she'd won a prize.

"You mean more of the same?" Meg asked. The idea of starting over was appalling, but she could do it. She'd learned you could do almost anything if it meant you might stay alive.

"No, been there, done that." Dr. Soames uttered the cliché in an embarrassed, self-conscious way. "What I mean is, chemo probably wouldn't yield the result we want."

"I thought I'd run out of treatments," Meg said. "I know a lot of women have radiation, but I didn't have any lymph node involvement." She almost laughed at the way she'd picked up the jargon. That's the way she'd done when computer programs were still mysteries to her.

"I'm suggesting a bone-marrow transplant," Dr. Soames said. "It's a risky treatment that involves using enough chemotherapy or radiation to kill off all cancer cells in your body."

"I thought I just had that," Meg said.

"We're talking about doses so high they wipe out your bone marrow," Dr. Soames said. "Marrow cells have to be reinjected so you can make blood cells again."

"It's called a bone-marrow transplant?"

"The popular name," Dr. Soames said. "The technical name is high-dose chemotherapy with stem-cell rescue."

"Does my insurance cover it?" Meg asked. She had decided instantly that she wanted the treatment. Risky or not, it seemed her only hope. The real question was whether she could afford it.

"Yours should," Dr. Soames said. "I've had other patients covered by the same company." He pressed his lips together. "And you'll need insurance. A bone-marrow transplant can cost two or three hundred thousand dollars."

Meg was stunned by the figure. But she was ready to fight, if her insurance company balked at paying. She knew that her life was at stake.

"You've been through a lot," Dr. Terrence said, giving Meg a sympathetic look. "Most of what you need to know about a marrow-cell transplant is in the consent document, so I hope you've had a chance to look at it."

"The hospital sent me a copy in Cleveland, and I read it three or four times," Meg said. "The form says a transplant *might* improve my chances of surviving. That sounds like a weak promise."

"Unfortunately, it's accurate," Dr. Terrence said. "We believe the treatment can help women who failed to get the full benefit from chemo, but we don't *know* that it can."

"I'm ready to do it anyway," Meg said. She sounded eager, as if welcoming the challenge.

"You don't need to decide now," Dr. Terrence said. "But I need to be sure that you understand the procedure and its risks before we can initiate treatment."

Bone-marrow transplantation, Dr. Terrence explained, was first used in the 1960s to treat leukemia. Drugs were used to wipe out the cancerous cells in the patient's bone marrow, but normal cells were also wiped out. These included the stem cells that produce blood cells and platelets. Patients without stem cells often died from the treatment, because their blood wouldn't clot and their immune system couldn't fight infection.

The early way of dealing with this was to inject the patient with marrow taken from a donor. If all went well, the stem cells from the donated marrow would reproduce and repopulate the patient's bone marrow. Because the donor cells were in a foreign body, however, they attacked the patient, causing a graft-versus-host reaction. The outcome was very often fatal.

"Improvements came along in the late sixties," Dr. Terrence said. "We don't use bone marrow anymore. We have machines that skim off normal stem cells circulating in a patient's blood. After we've wiped out her marrow, we inject her with these cells. They don't cause rejection, because they're her own cells."

"High-dose chemotherapy plus stem-cell rescue," Meg said. She sounded pleased with herself for recalling the name.

"Exactly," Dr. Terrence said, nodding.

"So if it works for leukemia, it should work for breast cancer?" Meg asked. She glanced at the flowers on the table, then looked back at Dr. Terrence. "I don't see the connection."

"Same logic," Dr. Terrence said. "If you use doses of chemicals high enough to kill the breast-cancer cells, you kill the stem cells in the bone marrow." He shrugged. "And that may kill the patient."

"Right," Meg said. She nodded. "If you can replace the stem cells lost from the bone marrow, you can get rid of the cancer and still keep the patient alive."

"That's the reasoning," Dr. Terrence said.

"And I don't need a bone-marrow donor?" Meg asked. "I might get my parents or one of my friends."

"Not necessary," Dr. Terrence said, shaking his head. "As I said, we can now harvest the stem cells from your blood."

"Cool," Meg said. She brushed a hand over her forehead as if pushing back her hair. "But why is this treatment riskier than regular chemo?"

"We use significantly higher doses of drugs," Dr. Terrence said. "So the drugs alone carry the risk of death. They can also cause nausea, vomiting, ulcers in the mouth and throat, and significant fatigue. Your immune system is severely compromised, so you may die from an infection. The transplanted stem cells might not take, so your immune system doesn't come back."

"Sort of extreme-chemo," Meg said. Her smile was strained. "I did okay with regular chemo."

"Then you know we can help with the side effects," Dr. Terrence said. "We can ease pain, but keep in mind that we can't always cure the infections. The mortality produced by the treatment itself is three to five percent."

"Just from the treatment, not the cancer?" Meg asked.

"Right," Dr. Terrence said. "Also, the quality of life during treatment is lower than for standard chemo. You have to stay in the hospital for weeks so we can keep an eye on you."

Meg folded her hands in her lap and focused her gaze on the flowers again. "But it's my best chance, isn't it?" she asked. She turned her head and looked at Dr. Terrence.

"That's not so clear," Dr. Terrence said. He spoke firmly, as if determined to forestall wishful thinking. "A few small studies show that the treatment offers a slight benefit, but none of the studies randomly assigned women to get either standard chemo or high-dose chemo plus stem cell rescue."

"So the studies could be wrong?" Meg frowned.

"Consider this," Dr. Terrence said. He learned forward in his chair. "Suppose the women who got standard chemo were mostly elderly, frail, and suffering from heart disease or diabetes, and suppose those who got chemo plus stem-cell transplants were younger, stronger, and generally healthy."

"Yeah, right," Meg said. She sighed. "If more of the younger and stronger ones survived, it might not be due to getting the new treatment. It might be because they were younger and stronger."

"Exactly," Dr. Terrence said. "We need controlled studies, and we don't have them yet."

"Dr. Soames definitely thought I should get a bone-marrow transplant," Meg said. "Is he way off base?"

"No, he's with the majority of breast-cancer oncologists," Dr. Terrence said. "But they're making a judgment call in the absence of genuine scientific evidence."

"And I've got to make one too," Meg said. She nibbled at her lower lip. "It's not as straightforward as I had expected. When Dr. Soames mentioned the treatment, I thought, sure it's risky, but it's my best bet, so I'm willing to go for it. Now you're saying it's not clear that it's my best bet?"

"I'm afraid so," Dr. Terrence said. "I wish we had the evidence now. People are doing the research, but you can't afford to wait for their results."

"Gads!" Meg said. She then unexpectedly smiled. "I'm used to people arguing about history and not being able to settle some question because the evidence is lacking. I didn't expect it in medicine, I guess."

"In history and science, you can suspend judgment and spend years waiting for more evidence," Dr. Terrence said. "But in medicine we don't have that luxury."

Meg turned her head slightly, then looked at Dr. Terrence out of the corners of her eyes. "Could I get you to say what you'd recommend if your wife or sister was in my situation?"

"I can tell you," Dr. Terrence said. "I'd recommend that they give serious thought to a stem-cell transplant, but I wouldn't go beyond that." He shook his head. "The evidence isn't there for me to say that the transplant is better than the standard treatment."

"Fair enough," Meg said. "Maybe I should just flip a coin."

"I can't recommend that," Dr. Terrence said. "Deciding here isn't like choosing between chocolate and vanilla. The transplant will make you much sicker and might even be fatal."

"But it might be effective," Meg said. "That's possible?"

"It's possible," Dr. Terrence admitted.

"Then I want to have the transplant," Meg said.

"You need to think about it more," Dr. Terrence said. "Read the consent document again, and see if you have more questions. Our protocol requires a twenty-four-hour minimum between the time of a meeting like this and your signing the consent form."

"I'm not going to change my mind," Meg said.

Once Meg signed the consent document, she was moved to an isolation unit in the hospital. The first step in her treatment was to harvest the stem cells from her bone marrow.

Meg was injected with colony-stimulating factor, a drug that boosts stem-cell production in the marrow. Her blood was then tested regularly over the course of seven days, and as soon as the number of circulating stem cells increased, she was hooked up by two IV lines to a machine a bit larger than a microwave oven.

One of the lines fed her blood into the machine, where a centrifuge spun it and separated the stem cells from the mature blood cells and platelets. The second IV returned the blood, minus the stem cells, to her body in a steady drip. The recovered stem cells were transferred to plastic tubes and frozen.

"It was a weird process," Meg later said. "It didn't hurt, really, but I found it strange to think of myself as a component in a circulatory system."

Harvesting the stem cells took only about four hours, and it was the easy part of Meg's treatment. The day after she donated the cells, she received the first round of the high-dose chemotherapy that Dr. Terrence had prescribed. The chemo was given to her in her room, rather than a chemotherapy suite. During that time, the medical staff and visitors had to scrub their hands for three minutes with bactericidal soap and put on paper gowns and shoe covers before entering her room. Precautions were taken with her meals to make sure that she didn't ingest bacteria from undercooked food or raw fruits and vegetables.

The chemo continued for three weeks.

"I was a lot sicker than I had been in Cleveland," Meg recalled after the treatment was over. "I would throw up until I had the dry heaves and thought my stomach was going to be turned inside out. I'd never experienced projectile vomiting before, but it happened every time. I thought about *The Exorcist*." She bared her teeth in a grimace.

"The nausea wasn't so bad, but the fatigue was worse than before. I could only lie in bed in a completely floppy state. I was like a doll that's lost its stuffing and is just an empty shape."

Even after Meg completed her chemo, she wasn't permitted to leave the hospital. Her immune system was now so severely crippled that even a slight cold could turn into a raging infection that drugs might not control. Remaining in isolation offered her the best chance to avoid bacteria and viruses.

"My memory of that period is mostly of nurses coming into my room in masks and gowns and sticking me with needles," Meg said later. She wrinkled her nose. "That seemed to happen every fifteen minutes at every hour of the day. I was so sick that I thought dying would have been perfectly all right."

She held up her right hand in a "stop" gesture. "Not that I wanted to die. I just didn't care if I lived." She took a deep breath and expelled it in a puff. "If that makes any sense."

Meg's blood wasn't drawn every fifteen minutes, but it was drawn six times a day on a regular schedule. Each time, the tube was sent to the hospital's lab, where it was tested for cancer cells. When, finally, the pathologists were unable to detect any abnormal cells in her blood, Meg was declared cancer-free. She was ready to receive the stem-cell transplant.

"I didn't react with great joy when I heard the cancer was gone," Meg recalled. "My feelings were so numb and I'd been so sick that all I cared about was getting through the day."

Dr. Terrence ordered one of the stored tubes of Meg's stem cells thawed and warmed. The cells were then mixed with a sterile solution and doses were dripped into her arm though an IV attached to a device that metered the amounts. Meg remained hooked up to the IV for the four days it took for her all the harvested stem cells to be returned to her blood.

Meg's blood, once again, was drawn regularly and tested. This time the aim was to determine if her stem cells had divided and repopulated her vacant bone marrow. If they had, they would then start producing normal blood cells and platelets.

The tests showed that they had. Meg was cancer-free and had a normally functioning immune system.

Meg had been confined to the hospital for almost six weeks. With her immune system functioning again, she was able to move out of the isolation ward. She spent the last few days of her stay in the same hospital room where I had met her.

I went with Dr. Terrence to see Meg on the day she was released. The view from the single window was no longer so dreary, because flowers were blooming on the small patch of ground at the center of the circular driveway.

Meg was significantly thinner. Her face was drawn and pale, and she had tied a blue bandanna around her head. She looked like someone who had been very ill. Yet her hazel eyes still shone with a greenish hue, and she continued to give the impression that at any moment she might laugh or at least smile. The hospital garb was gone. She was wearing tailored jeans and a fitted, long-sleeved white shirt.

"We'll miss you," Dr. Terrence said. Meg was sitting by her bed reading a paperback of *Night Vision*, and Dr. Terrence took the chair facing her. He smiled. "Not that I would want you to stay a minute longer than you need to."

"I'm surprised by how nervous I am," Meg said. She put the open book face-down on her bed. "I've been in the hospital so long that I worry about how I'll cope when I can't get help by pushing a call button."

"You'll do fine," Dr. Terrence said. His tone expressed no doubt. "Still planning to stay with your parents?"

"My mom is picking me up this afternoon," Meg said. "She wants to take care of me, and I'm feeling weak enough to be happy about it." She let her head drop, then she raised it and smiled. "Mostly, I just want to sleep."

"You're on the mend, though," Dr. Terrence said. "And in Silver Spring, you'll be close enough that if you have a problem, you can be back here in an hour."

"When will I know if the transplant worked?" Meg asked. Her brows drew together in a frown.

"Given that you're cancer-free and doing well, it's worked already," Dr. Terrence said. "But I think what you're really asking is, when can I be sure the cancer won't come back?"

"You're right," Meg said. "That's the last thought I have when I go to sleep and the first when I wake up."

"I wish I could tell you that it's never going to happen," Dr. Terrence said. "But I don't know if that's true." He paused, as if thinking about how to go on.

Meg's eyes remained fixed on him. She seemed to be waiting for him to pronounce a judgment.

"I don't know that the cancer *will* come back either," Dr. Terrence said. He shook his head. "We don't have the right data for me to give you even an estimate."

"So I have to learn to live with uncertainty," Meg said. She puffed up her cheeks, then blew the air out slowly.

"I'm afraid so," Dr. Terrence said. Then his tone became brighter. "But you can remind yourself that you got the best medical advice available and that you did everything you could to fight the disease."

"I gave it my best shot," Meg said, nodding. "I did." She lowered her eyes, and her face clouded. She seemed about to cry. Then she raised her head and smiled.

Despite everything Meg had been through, she looked wonderful. Medicine's magic seemed to have worked.

Neither Dr. Terrence nor I ever saw Meg again. Meg had returned to Cleveland and to her job after a few weeks at her parent's house. Dr. Terrence eventually received a formal report on her from Dr. Soames, her Cleveland oncologist.

The news wasn't good.

During one of Meg's routine exams, Dr. Soames discovered a mass in her abdomen, and a biopsy showed that her cancer had returned. This time Meg had no treatment options. No reputable physician believed that any therapy would push back the rising tide of multiplying cancer cells spreading throughout her body. Meg had, in the usual accusatory jargon, failed all therapies. The most any physician could do to help her was to give her psychological support and prescribe medications to help control her pain as she waited for her life to end.

Meg Anderson died in April 1995, eight months after her stem-cell transplant. She had been daring enough to take a chance on a dangerous and unproven therapy. Her doctors had offered it to her because they believed that it might be effective.

They were sadly and tragically mistaken.

Desperate diseases call for desperate measures.

Some oncologists in the early 1980s began to think it might make sense to extend a treatment established as effective in leukemia to patients with advanced breast cancer. After all, the problem with using chemotherapy to treat breast cancer is the same as with leukemia—doses high enough to kill off all the cancer cells will also destroy the blood-producing stem cells in the bone marrow and threaten the patient's life.

Hence, the reasoning went, just as in treating leukemia, a patient could be given massive doses of chemotherapy drugs, so long as this was followed by a stem-cell transplant. (I'll continue to call this a "bone-marrow transplant," because that remained the popular name even after the stem cells were acquired from the blood.) For the first time, doctors thought, patients with advanced breast caner would have a good chance to beat it. Given that less than 25 percent of women with stage IV cancer were alive in five years and only 10 percent were alive in ten years, almost any new treatment that made sense seemed worth the gamble.

STANDARD OF CARE—ALMOST

The women oncologists considered the best candidates were those who were younger, healthier, and under fifty. The treatment was so harsh that it had a 3–5 percent mortality, and women who were physically strong had the best chance of surviving. Also, younger women often have young children depending on them. Finally, younger women have their productive lives ahead of them, so the lost years deprive them of opportunities. Thus, most physicians felt that not only were younger women the best candidates for a BMT, they deserved a fighting chance to save their lives.

Insurance companies weren't enthusiastic. A BMT might cost $200,000–$300,000, and most insurers refused to pay for one on the grounds that BMTs were "experimental." A large number of women who believed their insurance companies were denying them their only chance to go on living filed law suits to force the companies to pay. Some insurers settled out of court, but they didn't change their practice of denying coverage for BMTs. A few insurers consulted medical experts and decided to pay for BMTs after standard therapies had failed.

Many patients thought they might be dead before their suits forced their insurance companies to pay. Women with resources paid with their own money, and others borrowed the money from their families and friends. Some sold their houses, and a few made public appeals for contributions to allow them to get the treatment they believed they needed.

The conflicts between insurers and women wanting a BMT coincided with the growth of the women's movement and the recognition of breast cancer as a major health problem. Thus, BMTs became identified as a "women's issue," and advocates often championed individual cases and pressed insurers and legislators to guarantee women access to the treatment. Several states passed laws requiring insurance companies to pay for BMTs.

By the end of the 1990s, a BMT was viewed almost as a standard therapy for a select group of women, yet its effectiveness still hadn't been established. A few small clinical trials were conducted, but they weren't scientifically rigorous. They included women of different ages and physical conditions, and they didn't randomly assign patients to the experimental or the standard treatment groups. The confidence in BMTs by oncologists continued to be based mostly on the analogy with treating leukemia.

PUT TO THE TEST

Clinical trials comparing the effectiveness of the standard treatment with a BMT began in 1988, but they weren't completed for eleven years. They dragged on because of the difficulty in enrolling enough patients to get statistically meaningful results. Oncologists had more confidence in BMTs, so they didn't want their patients to participate in studies in which they might be randomly assigned to get the standard treatment.

The data from four well-conducted studies were released in 1999. They showed no difference in the effectiveness of standard chemotherapy compared with high-dose chemotherapy followed by a stem-cell transplant. Thus, the dramatic, expensive, grueling, and risky procedure that had been the last, best hope of so many women with late-stage breast cancer turned out to be no better than another round of chemotherapy. Hospitals stopped offering the procedure, and oncologists stopped using it. Insurance companies that had paid for BMTs informed physicians and patients that they would no longer cover the treatment.

HINDSIGHT

BMTs most likely shortened the lives of many women. With its mortality rate ranging from 5 percent to 15 percent in the early days of its use to about 3 percent at the end, the treatment probably caused the deaths of many women who might have survived longer had they received standard chemotherapy. Even the women who survived the treatment experienced much unnecessary suffering without gaining any extra benefit.

Women unable to get their insurers to cover a BMT and unable to raise the $100,000–200,000 they needed to pay for one were left with feelings of frustration, anger, and despair. Many had come to believe that a BMT was their only hope of staying alive, yet they were powerless to get one. Women who raised the money needed for a BMT often did so by such extraordinary measures as exhausting the savings of their families, selling their houses, or going deeply into debt. Thus, many people ended up ruining themselves financially to pay for a treatment that offered no special advantage.

From the social perspective, the money was also poorly spent. The $100,000,000 or more used to pay for BMTs over the years could have been better used. Patients could have received the standard treatment and been no worse for it, and the additional money could have paid for the care of those with other diseases or gone to support medical research. The money spent on an unproved and relatively ineffective treatment was money wasted.

MEG ANDERSON

Meg Anderson had late-stage breast cancer in 1995. The standard regimen of lumpectomy and chemotherapy had left her with cancer—a tumor in her lung and, most likely, cancer cells elsewhere in her body. Dr. Soames, her oncologist, considered Meg an excellent candidate for a BMT. She was young and strong, with good general health and, except for her cancer, many years of life ahead of her.

Dr. Terrence told Meg that stem-cell transplants hadn't been shown to be successful in treating her disease and that the results of the small studies that had been done weren't reliable. Meg was also given an informed-consent document that described the treatment and its risks, mentioned its severe side effects (including death), informed her that the treatment was not proven, and advised her that she might gain no benefit from it. Dr. Terrence answered her questions and allowed her a sufficient amount of time to deliberate. Meg's decision to have the treatment was an expression of her autonomy, and by any reasonable standard, the conditions for informed consent were satisfied.

I suggest that Meg didn't make a bad decision. What went wrong for her and hundreds of other women was the prevailing climate of opinion when they had to make a decision. So many physicians had come to accept BMTs as an effective treatment that Meg Anderson might be thought foolish, cowardly, or even irrational had she *not* chosen a BMT.

Consider these factors: Dr. Soames told her she was good candidate and that a BMT was her best option; Meg very much wanted to live; she was willing to undergo the suffering and risks required by the treatment; her insurance company would pay for it; major academic medical centers offered it; no one with medical expertise advised her not to have a BMT, even though no one told her it would work.

Who wouldn't have decided as Meg did? Maybe someone not so passionately committed to life or someone with a greater fear of suffering.

The many factors in the situation favoring choosing a BMT weren't outweighed by the fact that evidence wasn't available to prove it was effective. Stem-cell transplants had been so widely and uncritically accepted that oncologists tended to think of them as a conventional therapy that would soon be validated by clinical trials. The treatment seemed so reasonable on theoretical grounds that waiting for the scientific evidence seemed only a formality.

The lesson for medicine here is that it must remain vigilant and not allow a treatment, no matter how prima facie reasonable it appears, to gain the status of a standard therapy without clinical testing. Patients tend to think that any novel therapy is better than a relatively ineffective standard one, and when physicians seem enthusiastic about it, for desperate patients that's equivalent to an endorsement.

The lesson for patients is to guard against the impulse to believe that an experimental treatment is more effective than a tested one, even if the tested one seems of limited value. Would Meg Anderson have lived longer had she rejected a BMT and opted for more chemotherapy? Perhaps not. She would have been spared the greater unpleasantness of a BMT, but maybe the feeling of hope that it gave her offset the nausea. When no treatment can be counted on to prolong life, we look for small things to make it satisfactory.

Meg Anderson made the best choice she could—given the prevailing climate of opinion.

Acknowledgments

I thank my colleagues Richard Cook and Richard Wright for their encouragement and my friend Ted Trimble for arranging for me to spend time at Johns Hopkins School of Medicine. I am particularly grateful to Peter Prescott, my editor at Oxford University Press. Without his enthusiastic support, this book may never have been completed. I am indebted to Keith Faivre, my production editor, for his extra work, good humor, and welcome expertise.

Notes

Preface

Frank Huyler's *The Blood of Strangers: Stories from Emergency Medicine* (Berkeley: University of California Press, 1999) was my model for using sharp, clear language to capture the intimate experiences of people facing serious medical problems. The book deserves to be more widely known.

Huyler sees the world from the perspective of an ER doctor; Atul Gawande sees it from the point of view of a surgeon. Eloquent in its own way, Gawande's *Complications* (New York: Metropolitan Books, 2002) shows us how surgical errors can occur despite appropriate caution and the best of intentions. He lets us see that judgment also requires training.

Robert Lipsyte's *In the Country of Illness: Comfort and Advice for the Journey* (New York: Knopf, 1998) conveys the fear, confusion, and tedium inherent in being a patient undergoing lengthy treatment for a life-threatening illness (testicular cancer, in his case). The necessity of winning the favor of the indifferent, often self-absorbed administrative assistants who are the guardians of appointment times and specialist referrals is a detail never mentioned in books by physicians. Lance Armstrong (with Sally Jenkins) tells what it's like to be a celebrity patient with the same disease as Lipsyte in *It's Not About the Bike: My Journey Back to Life* (New York: Putnam, 2000).

Perhaps immodestly, I would like to suggest my book *Intervention and Reflection: Basic Issues in Medical Ethics* (Belmont, CA: Thomson Wadsworth, 2008) as a useful introduction to the wide range and great variety of problems in recent medical ethics and bioethics. Half the book is devoted to presenting cases in medical ethics, sketching the contexts in which problems arise, summarizing relevant scientific and medical information, and discussing ethical issues and ways of resolving them. The last section of the book provides brief accounts of basic ethical theories and principles and discusses ways they can be used in making and justifying decisions about ethical issues arising in medicine. On this topic I also suggest John D. Arras, "The Way We Reason Now: Reflective Equilibrium in Bioethics," in Bonnie Steinbock, ed., *Oxford Handbook of Bioethics* (Oxford: Oxford University Press, 2007), pp. 46–71.

My account of the Quinlan case is based on Joseph and Julia Quinlan and Phillis Battelle's *Karen Ann Quinlan* (New York: Doubleday, 1977); B. D. Colen, *Karen Ann Quinlan: Dying in the Age of Eternal Life* (New York: Nash, 1976); and *In the Matter of Karen Quinlan: The Complete Legal Briefs, Court Proceedings, and Decisions* (Arlington, VA: University Publications of America, 1975; no author or editor listed). For a fuller review of the issues raised by the Quinlan case, see "Euthanasia and Physician Suicide" in my book *Intervention and Reflection*. The chapter also contains accounts of the cases of Nancy Cruzan and Terri Schiavo. Brian Clark's 1978 play *Whose Life is It Anyway?* was originally written as a television play and was produced in 1972. A movie version, also written by Clark and starring Richard Dreyfuss and John Cassavetes, was released in 1981.

Chapter One: The Woman Who Decided to Die

For a sophisticated analysis of the concept of autonomy, including its use in medical ethics, see Bruce Jennings, "Autonomy," in Bonnie Steinbock, ed., *Oxford Handbook of Bioethics* (Oxford: Oxford University Press, 2007), pp. 72–90.

Michael Gearin-Tosh's *Living Proof: A Medical Mutiny* (New York: Scribner, 2002) is an autobiographical account of an Oxford professor who was diagnosed with cancer at age fifty-four but who refused to take the advice of several specialists who urged him to seek treatment without delay. Gearin-Tosh exercised his autonomy at the cost of increasing his risk of dying by delaying treatment. Katrina Firlik's *Another Day in the Frontal Lobe: A Brain Surgeon Exposes Life on the Inside* (New York: Random House, 2006) is a neurosurgeon's account of her experiences in helping patients and their families make decisions about whether additional treatment might be useful. The book is more thoughtful than its title and subtitle suggest.

Chapter Two: Like Leaving a Note

For a discussion of organ donation, altruism, organ sales, and criteria for establishing death, I would like to suggest my book *Raising the Dead: Organ Transplants, Ethics, and Society* (New York: Oxford University Press, 2002). Particularly relevant to the issues raised by this case are chapter 4, "That Others May Live"; chapter 5, "Kidney for Sale"; and chapter 8, "But Are They Really Dead?"

Charles R. Morris, in chapter 4 ("The Most Precious Resource") of his book *The Surgeons: Life and Death in a Top Heart Center* (New York: Norton, 2007), offers a detailed, vivid, and accurate account of how the transplant system works in acquiring and distributing organs and how a heart is transplanted.

Statistical information about the number of people on transplant waiting lists, waiting times, deaths while waiting, donors, transplants performed per year, survival times, and success rates can be found on the Web site of the United Network for Organ Sharing (UNOS): www.unos.org. It also provides access to information about the policies that govern the acquisition and distribution of transplant organs. The statistics I use are from UNOS.

Chapter Three: The Agents

The standard criteria for diagnosing mental disorders are those in the fourth edition of the *Diagnostic and Statistical Manual of Mental Disorders* (New York: American Psychiatric Association, 1994). This book is always referred to as *DSM-IV.* To learn more about the difficulty of determining when someone who is mentally ill (particularly someone with schizophrenia) is unable to exercise autonomy, see Jeanette Kennett, "Mental Disorder, Moral Agency, and the Self," in Bonnie Steinbock, ed., *Oxford Handbook of Bioethics* (Oxford: Oxford University Press, 2007), pp. 90–113.

Claire and Mia Fontaine's *Comeback: A Mother and Daughter's Journey Through Hell and Back* (New York: Reagan Books, 2006) is a vivid account of a struggle involving drug addiction, cutting behavior, and involuntary civil commitment. Mia Fontaine, the daughter, argued in a *New York Times* opinion piece ("Learning From Britney's Trouble," February 10, 2008) in favor a stronger policy of involuntary commitment than is now possible under laws written to offer extensive protection to civil liberties. Fontaine claims that because of these laws, parents must often stand by helplessly while a hospital frees their mentally ill or drug-addicted children. She suggests that the troubled singer Britney Spears might have received the help she needed sooner had the laws made it possible.

Susana Kaysen's *Girl, Interrupted* (New York: Turtle Bay Books, 1993) gives an autobiographical account of what it's like to be an eighteen-year-old who feels the need to step out of the ordinary world and seek asylum in an institution. She sought refuge at the hospital where Nancy Trail was sent. Roy Porter's *Madness: A Brief History* (New York: Oxford University Press, 2002) is a wide-ranging, informative, and readable account of the ways we have identified, explained, and treated mental illness in the Western world.

Chapter Four: Unsuitable

The shortage of transplant organs and the need for living-donor transplants is dealt with in my chapter ("Organ Transplantation") in Bonnie Steinbock, ed., *Oxford Handbook of Bioethics* (Oxford: Oxford University Press, 2007), pp. 211–239. The chapter discusses, among other topics, whether living donors should be permitted, the potential for coercion, benefits to recipients vs. risks to donors, strangers as donors, "heroic" donors, paying donors, and making commitments to living donors legally enforceable. R. W. Evans, et al., "The Potential Supply of Organ Donors" (*JAMA* 1992, 259: 1546–1547) offers convincing evidence that the supply of organs from deceased donors will never meet transplant needs. The specific ethical issues surrounding potential liver-segment donors are discussed in P. A. Singer, et al., "Ethics of Liver Transplantation with Living Donors" (*New England Journal of Medicine* 1989, 321: 620–622).

One of the best cases made for organ sales is Janet Radcliffe-Richards, et al., "The Case for Allowing Kidney Sales" (*Lancet* 1998, 351: 1950–1952). For divergent views, see J. B. Dossetor and V. Manickavel, "Commercialization: The Buying and Selling of Kidneys," in C. M. Kjellstrand and J. B. Dossetor, eds., *Ethical Problems in Dialysis and Transplantation* (Netherlands: Kluwer Academic Publishers, 1992), pp. 61–71. The issues raised by selling organs are also usefully discussed in Carl Cohen, "Public Policy and the Sale of Human Organs," *Kennedy Institute of Ethics Journal* (2002, 12: 47–64).

Chapter Five: Nothing Personal

For basic information about clinical trials, see the account in chapter 1 ("Research Ethics and Informed Consent") in my book *Intervention and Reflection: Basic Issues in Medical Ethics*, 8th ed. (Belmont, CA: Thomson Wadsworth, 2008). This chapter also presents several cases about which serious ethical questions have been raised, such as Jesse Gelsinger's death in a gene-therapy trial, Cold War radiation experiments, and the Willowbrook Hepatitis Experiment. Chapter 3 ("Race, Gender and Medicine") presents the infamous case of the Tuskegee Syphilis Study.

Current policies governing the federal regulation of clinical trials can be found at http//clinicaltrials.gov. The "Frequently Asked Questions" section explains how people might qualify for access to an experimental drug without being in a clinical trial. The Food and Drug Administration, in effect, permits drug manufacturers to initiate an uncontrolled observational study of patients taking the drug. These patients are treated under a special investigational drug protocol. This is possible, as I mentioned in presenting the case, only when the patient cannot qualify for the clinical trial for some reason such as age or poor physical condition, and when the standard therapy is believed not to be effective.

The basic principles governing all U.S. human research are presented in *The Belmont Report: Ethical Principles and Guidelines for the Protection of the Human Subjects of Research* (Washington, DC: National Commission for the Protection of Human Subjects of Biomedical and Behavioral Research, Department of Health, Education, and Welfare; April 18, 1979, http:/ohsr.od.nih.gov/guidelines/belmont.html). This document is always called the *Belmont Report*.

Chapter Six: "He's Had Enough"

For a discussion of the issues that arise in end-of-life situations, see chapter 11 ("Euthanasia and Physician-Assisted Suicide") in my book *Intervention and Reflection: Basic Issues in Medical Ethics*, 8th ed. (Belmont, CA: Thomson Wadsworth, 2008). The chapter presents the cases of Nancy Cruzan and Terry Schiavo, and it reprints an excerpt from the New Jersey Supreme Court decision in the Quinlan case. This is particularly relevant to the issue that faced Martha Post: the need to decide for someone who is no longer competent. The articles that follow the introductory materials in the chapter represent the variety of current thinking on these topics. Also worth close attention is James Rachels, "Active and Passive Euthanasia," pp. 725–729.

The sadness and confusion of what it's like to have a spouse die is powerfully evoked in Joan Didion's *The Year of Magical Thinking* (New York: Knopf, 2005) and, very differently, in Calvin Trillin's *About Alice* (New York: Random House, 2006).

Chapter Seven: Not More Equal

The case of a California prison inmate who received a heart transplant likely to cost the state a million dollars in overall medical care stirred considerable controversy in 2002. For a report on the case, see James Steingold, "Inmate's Transplant Prompts Questions of Costs and Ethics" (*New York Times*, January 31, 2002), and "Change of Heart" on *Sixty Minutes* (September 14, 2003). The prisoner, convicted of armed robbery, received the heart at Stanford University Medical Center. Not all states have provided prisoners with needed organ transplants. Arkansas, for example, refused to provide a liver transplant to James Earl Ray, the convicted killer of Martin Luther King Jr.

The Ethics Committee of the United Network for Organ Sharing states the organization's view in "Position Statement Regarding Convicted Criminals and Transplant Evaluation" (unos.org/resources/bioethics). An illuminating debate is presented in Martin F. McKneally and Robert M. Sade, "The Prisoner Dilemma: Should Convicted Felons Have the Same Access to Heart Transplants as Ordinary Citizens?" (*Journal of Thoracic and Cardiovascular Surgery*, 2003, 125: 451–453). For a discussion of the general issue of allocating organs on the basis of social-worth criteria, see chapter 3 ("Mickey Mantle's Liver Part II: The Issues") in my book *Raising the Dead: Organ Transplants, Ethics, and Society* (New York: Oxford University Press, 2002).

Chapter Eight: The Last Thing You Can Do for Him

William Bonadio's *Julia's Mother: Life Lessons in the Pediatric ER* (New York: St. Martin's, 2000) is a sensitive account of the struggles, triumphs, and failures in the emergency room of a children's hospital. For my account of how the criteria for brain death are applied and how a persistent vegetative state is diagnosed, I'm indebted to Kenneth V. Iserson's *Death to Dust: What Happens to Dead Bodies*

(Tucson, AZ: Galen Press, 1994), pp. 18–20. This endlessly fascinating, somewhat macabre book takes a wide view of its subject. The book has a question-and-answer format: How do nurses prepare bodies for the morgue? What do undertakers do? Its appendices include useful documents like the Universal Determination of Death Act.

The Harvard Ad Hoc Committee's arguments for the need for a new definition of death, and the definition itself, can be found in "Report of the Ad Hoc Committee of the Harvard Medical School to Examine the Definition of Brain Death" (*JAMA* 1968, 205: 337–340). The continuing debate about criteria for determining death is represented well in Stuart J. Youngner, Robert M. Arnold, and Renie Schapiro, eds., *The Definition of Death: Contemporary Controversies* (Baltimore, MD: Johns Hopkins University Press, 1999). The editors' introduction puts the issues in perspective. Fred Plum's account of diagnosing death is useful for appreciating the debates, and Martin Pernick's account of the workings of the Harvard Ad Hoc Committee identifies the agendas of some of the members.

Chapter Nine: The Boy Who Was Addicted to Pain

General information about assessment of people who may be delusional, suicidal, or in need of protection is presented in section 194, "Psychiatric Emergencies," in Mark H. Beers and Robert Berkow, eds., *Merck Manual of Diagnosis and Therapy* (Whitehouse Station, NJ: Merck Research Laboratories, 1999). See p. 1574 for advice about involuntary civil commitment.

As with most psychiatric disorders, the cause of cutting isn't completely understood. It can be a symptom of borderline-personality disorder, a major and almost intractable psychiatric disorder; but it can also be associated with a number of other conditions, including depression, child abuse or neglect, social isolation, post-traumatic stress disorder, or the refusal of an adolescent's parents to allow their child to express negative feelings. Fictional accounts of cutters usually portray them as adolescent girls or young women, but some evidence indicates that boys and young men are just as likely to be cutters. Psychiatric treatments are as varied as the putative causes. Antidepressants and psychotherapy are typical, and some psychiatrists think that cognitive therapy—getting cutters to understand why they cut themselves—is the most effective approach. Some also believe that group therapy is more effective than individual counseling.

Claire and Mia Fontaine's *Comeback: A Mother and Daughter's Journey Through Hell and Back* (New York: Reagan Books, 2006) tells the story of Claire's attempt to rescue Mia, her fifteen-year-old daughter, from her life as a high-school dropout, heroin addict, and cutter. Mia, unlike Chase Granger, was hospitalized involuntarily, although not for long enough to get the help she needed.

Chapter Ten: It Seemed Like a Good Idea

An excellent guide to breast cancer and its treatment is the American Cancer Society's *Breast Cancer Clear and Simple: All Your Questions Answered* (Atlanta, GA: American Cancer Society, 2008). The study results showing the comparative lack of effectiveness of bone marrow transplants were presented at the May 15, 1999, meeting of the American Society of Clinical Oncology and summarized in "Mixed Results in High-Dose Chemotherapy/Bone Marrow Transplant Studies for Women with Breast Cancer" at www.asco.org. The effectiveness of BMTs continues to be debated in medical journals, particularly in the *Journal of Clinical Oncology*. For responses to the 1995 report, see Denise Grady, "Doubts Raised on a Breast Cancer Procedure" (*New York Times*, April 16, 1999).

The statistics comparing BMTs to conventional therapy in 1995 are reported in Susan M. Love (with Karen Lindsey), *Dr. Susan Love's Breast Book* (New York: Addison-Wesley, 1995). Barron H. Lerner's *Breast Cancer Wars* (New York: Oxford University Press, 2001) is an excellent survey of the debates about breast cancer treatments, particularly about whether lumpectomy is as effective as mastectomy. David Biro's *One Hundred Days: My Unexpected Journey from Doctor to Patient* (New York: Pantheon, 2000) is the compelling autobiographical account of a young physician diagnosed with a rare blood disease who, despite his profession, experiences the anxieties, confusions, and doubts characteristic of other patients. He is treated with a BMT using cells from donor marrow.

Journalist Geralyn Lucas's *Why I Wore Lipstick to My Mastectomy* (New York: St. Martins, 2004) paints a vivid picture of how much a woman's life changes when she is diagnosed and treated for breast cancer. *Breast Cancer? Let Me Check My Schedule*, by Peggy McCarthy and Jo An Loren, eds. (Boulder, CO: Harper-Collins Publishers, 1997), charts the experiences of women who face diagnosis, treatment, and the recurrence of their disease.

An earlier version of this chapter was written in collaboration with Edward L. Trimble, M.D.

About the Author

Ronald Munson, a nationally recognized bioethicist, is Professor of the Philosophy of Science and Medicine at the University of Missouri–St. Louis. He received his Ph.D. from Columbia University and was a Postdoctoral Fellow at Harvard University. He has been a Visiting Professor at the University of California–San Diego, Johns Hopkins School of Medicine, and Harvard Medical School.

Munson is a medical ethicist for the National Eye Institute and a consultant for the National Cancer Institute. He is also a member of the Washington University School of Medicine's Human Research Protection Committee and an Associate Editor of the *American Journal of Surgery*.

His other books include *Raising the Dead: Organ Transplants, Ethics, and Society* (Oxford University Press, 2002), named a "Best Book in Science and Medicine" by the American Library Association, and *Reasoning in Medicine* (with Daniel Albert and Michael Resnick). His *Intervention and Reflection: Basic Issues in Medical Ethics*, in print for thirty years and now in its eighth edition, is the most widely used bioethics text in the United States. Munson is also the author of the novels *Nothing Human*, *Fan Mail*, and *Night Vision*.